C005058577 0

Exponential

Exponential

How Accelerating Technology Is Leaving Us Behind and What to Do About It

Azeem Azhar

1 3 5 7 9 10 8 6 4 2

Random House Business
20 Vauxhall Bridge Road
London SW1V 2SA

Random House Business is part of the Penguin Random House group of companies whose addresses can be found at global.penguinrandomhouse.com.

First published in the United Kingdom by Random House Business in 2021

www.penguin.co.uk

A CIP catalogue record for this book is available from the British Library.

ISBN 9781847942906 (hardback)
ISBN 9781847942913 (trade paperback)

Typeset in 11.5/15.5pt Sabon MT Std by Jouve (UK), Milton Keynes
Printed and bound in Great Britain by Clays Ltd, Elcograf S.p.A.

The authorised representative in the EEA is Penguin Random House Ireland, Morrison Chambers, 32 Nassau Street, Dublin D02 YH68

Penguin Random House is committed to a sustainable future for our business, our readers and our planet. This book is made from Forest Stewardship Council® certified paper.

To Salman, Sophie and Jasmine
and the *Exponential View* community

Contents

Preface: The Great Transition

My home lies between the neighbourhoods of Cricklewood and Golders Green in north-west London. It is a suburban house in a suburban street of a type familiar across Europe and the United States. And it is a relatively recent addition to the landscape. Look at a map of the area from 1920, and all you'll see is farmland. The plot of my semi-detached house is right in the middle of a field. A bridleway is shown where an access road now runs, and a few gates and hedgerows demarcate what is now my neighbourhood. A couple of hundred metres to the north lies a blacksmith's.

Just a few years later, the area had transformed. Pick up a map of the same area from 1936 and you'll see the farmland has become the streets that I walk through daily. The blacksmith's has disappeared, replaced by a mechanical workshop. The brick-built interwar-era homes are arranged on the same plots they occupy now, perhaps lacking the odd glass extension. It is a remarkable metamorphosis, which reflects the emergence of a recognisably modern way of life.

As late as the 1880s, life in London resembled that of a much earlier era – horses plied the roads, leaving piles of manure in the streets as they did so; most domestic tasks were powered by hand; much of the population inhabited crowded, centuries-old slum buildings. But beginning in the 1890s, and in many cases completed by the 1920s, the key technologies of the twentieth century took hold. Pictures of central London streets in 1925 show them free of horses, replaced by cars and buses. A network of cables would have carried electricity from coal-fired power stations to offices and homes. Telephone lines ran into many houses and allowed people to talk to distant friends.

These changes in turn brought social upheaval. As modern systems of production developed, so too did full-time employment contracts with benefits; new forms of transport brought with them the commute; the electrification of factories helped the rise of large companies with recognisable brand names. Someone living in the 1980s who stepped into a time machine and went back to the 1880s would have seen little they were familiar with. If they travelled back to just the 1930s they would have recognised much more.

This two-decade transformation reflects the sudden, dramatic changes that technology can bring. Since the days of flint axes and wooden digging sticks, humans have been technologists. We seek to make life easier for ourselves; and to do so, we build tools – technologies – that help us achieve our goals. These technologies have long allowed humans to redefine the world around us. They let us farm and then build; travel on land, then through air, then into space; move from nomadic life to villages to cities.

But, as my predecessors in what is now north-west London learnt, the technologies we build can take society in unexpected directions. When a technology takes off, its effects can be enormous, stretching across all the areas of human life: our jobs, the wars we fight, the nature of our politics, even our manners and habits. To borrow a word from economics, technology is not 'exogenous' to the other forces that define our lives – it combines with political, cultural and social systems, often in dramatic and unforeseen ways.

The unpredictable ways that technology combines with wider forces – sometimes moving slowly, sometimes causing rapid and seismic transformations – are what makes it so difficult to analyse. The emerging discipline of complexity science tries to make sense of the ways in which the different elements of a complicated system interact – how different species relate to each other to make up an ecosystem, for example. Human society is the ultimate 'complex system'; it is made up of countless, constantly interacting elements – individuals, households, governments, companies, beliefs, technologies.

According to complexity science, the connections between different elements mean that small changes in one area of a system can ripple across the whole. And these changes can be chaotic, sudden and profound.[1] Even if we have a significant degree of knowledge about the component parts of the system, establishing where these ripples might end up is rarely straightforward.[2] A new technology might at first cause a small social change – but one that eventually spirals into major repercussions for the whole of society.

When these ripples – or 'feedback loops', in the jargon of complexity science – start to spread, they can feel uncomfortable. One need only glance at the pages of a newspaper from the turn of the twentieth century to realise that sudden change is anxiety-inducing. A quick survey of *New York Times* articles from a century ago reveals that Americans were apprehensive about elevators, the telephone, the television and more.[3]

Of course, jitters in the elevator were rarely the real issue. Rather, these innovations came to symbolise people's fears about the pace of change. We know intuitively that technological changes rarely remain enclosed within one sphere. By allowing us to build ever-taller buildings, elevators revolutionised the layout and economies of cities. By making contact between people easier, the telephone drastically altered how humans interacted with colleagues and friends. After a technology has taken off, its effects are felt everywhere.

Today, we are undergoing another period of dramatic transformation. The clearest sign of this is the way people talk about technology. The PR company Edelman runs a renowned annual survey on trust in the public sphere. One of their key questions – put to 30,000 people in 20 countries – is whether they feel comfortable with how quickly technology was moving. In 2020, more than 60 per cent of respondents felt the 'pace of change was too fast', a number that had been creeping upwards for several years.[4]

It's tempting to assume that people always feel technological and social change is too fast. They thought so a century ago, and they think so again now. But the argument of this book is that we are indeed living through a time of unusually fast change – and this change is being brought about by sudden technological advances. In the early twenty-first century, the defining technologies of the industrial age are metamorphosing. Our society is being propelled forward by several new innovations – computing and artificial intelligence, renewable electricity and energy storage, breakthroughs in biology and manufacturing.

These innovations are improving in ways that we don't yet fully understand. What makes them unique is the fact they are developing: at an exponential pace, getting faster and faster with each passing month. As in previous periods of rapid technological change, their impact is felt across society – not only leading to new services and products, but also altering the relationship between old and new companies, employers and workers, city and country, citizens and the market.

Complexity scientists refer to moments of radical change within a system as a 'phase transition'.[5] When liquid water turns into steam, it is the same chemical, yet its behaviour is radically different. Societies too can undergo phase changes. Some moments feel abrupt, discontinuous, world-changing. Think of the arrival of Columbus in the Americas, or the fall of the Berlin Wall.

The rapid reorganisation of our society today is just such a moment. A phase transition has been reached, and we are witnessing our systems transforming before our very eyes. Water is becoming steam.

The transformation of society in the early twenty-first century is the focus of this book. It is a book about how new technology is getting faster. And it tries to explain the effect this acceleration is having on our politics, our economies and our ways of life.

But it is not a pessimistic book. There is nothing inevitably harmful about the technologies I will describe. The elements of society that are most important to us – our companies, cultures and laws – emerged in response to the changes brought by earlier technologies. One of the defining features of human history is our adaptability. When rapid technological change arrives, it first brings turmoil, then people adapt, and then eventually, we learn to thrive.

Yet I have chosen to write this book because we presently lack the vocabulary to make sense of technological change. When you watch the news, or read the blogs coming out of the longstanding capital of tech, Silicon Valley, it becomes apparent that our public conversation around technology is limited. New technology is changing the world, and yet misunderstandings about what this tech is, why it matters and how we should respond are everywhere.

In my view, there are two main problems with our conversation about technology – problems that this book hopes to address. First, there is a misconception of how humans relate to technology. We often assume that tech is somehow independent of humanity – that it is a force that brought itself into being and doesn't reflect the biases and power structures of the humans who created it. In this rendering, technology is value-free – it is made neutral – and it is the consumers of the technology who determine whether it is used for good or for evil.

This view is particularly common in Silicon Valley. In 2013, Google's executive chairman Eric Schmidt wrote, 'The central truth of the technology industry – that technology is neutral but people are not – will periodically be lost amid all the noise.'[6] Peter Diamandis, an engineer and physician – as well as the founder of a company that offers tech courses, Singularity University – wrote that while the computer 'is clearly the greatest tool for self-empowerment we've yet seen, it's still only a tool, and, like all tools, is fundamentally neutral'.[7]

This is a convenient notion for those who create technology. If technology is neutral, its inventors can concentrate on building their gizmos. If the tech starts to have any insidious effects, society – rather than its inventors – is to blame. But if technology isn't neutral – that is, if it has encoded some form of ideology, or system of power – that might mean its makers need to be more careful. Society might want to manage or regulate the technologists and their creations more carefully. And those regulations might become a hassle.

Sadly for these engineers, their view of technology is a fiction. Technologies are not just neutral tools to be applied (or misapplied) by their users. They are artefacts built by people. And these people direct and design their inventions according to their own preferences. Just as some religious texts say that humans are fashioned in the image of God, so tools are made in the image of the humans who design them. And that means our technologies often recreate the systems of power that exist in the rest of society. Our phones are designed to fit in men's hands rather than women's. Many medicines are less effective on Black and Asian people, because the pharmaceutical industry often develops its treatments for white customers. When we build technology, we might make these systems of power more durable – by encoding them into infrastructure that is more inscrutable and less accountable than humans are.

And so this book doesn't analyse technology as some abstract force, separate from the rest of society. It views tech as something that is built by humans and reflects human desires, even if it can also transform human society in radical and unexpected ways. *Exponential* is as much about the way technology interacts with our forms of social, political and economic organisation as it is about the technology itself.

The second problem with the way we talk about technology is even more insidious. Many people outside of the world of technology make no effort to understand it, nor to develop the right response to it. Politicians frequently demonstrate a fundamental ignorance about even the most basic workings of mainstream technologies.[8] They are like people trying to fuel a car by filling its trunk with hay. The Brexit trade deal, agreed between the UK and the European Union in December 2020, describes the Netscape Communicator as a 'modern e-mail software package'. The software has been defunct since 1997.

Admittedly, understanding new technology is hard. It takes knowledge of a wide range of new innovations. And it also takes an understanding of society's existing rules, norms, institutions and conventions. In other words, effective analysis of technology involves straddling two worlds. It's reminiscent of a famous 1959 lecture by the British scientist and novelist C. P. Snow. He feared that intellectual life was being split between the domains of literature and science, specifically within the context of British public life. These 'two cultures' did not intersect, and those who understood one rarely understood the other – there was a 'gulf of mutual incomprehension', generated by a 'backward looking intelligentsia' made up of artsy Oxbridge graduates who looked down on technological and scientific progress. This, according to Snow, had disastrous implications: 'When those two senses have grown apart, then no society is going to be able to think with wisdom.'[9]

Today, the gap between the two cultures is wider than ever before. Except now it is most pronounced between technologists – whether software engineers, product developers or Silicon Valley executives – and everyone else. The culture of technology is constantly developing in new, dangerous and unexpected directions. The other culture – the world of humanities and social science, inhabited by most commentators and policymakers – cannot keep track of what is happening. In the absence of a dialogue between the two cultures, our leading thinkers on both sides will struggle to offer the right solutions.

This book is my attempt to bring these two worlds together. On the one hand, I will try to help technologists view their efforts in a wider social context. On the other, I'll aim to help non-technologists get a better understanding of the technologies underpinning this period of rapid social change.

This mix of disciplines suits me well. I am a child of the microchip, born the year after the first commercially produced computer processor was released; a young adult of the internet, who discovered the web while at university; and a professional of the tech industry, having launched my first website – for Britain's *Guardian* newspaper – in 1995. I have founded four tech companies and invested in more than 30 start-ups since 1998. I even survived the dot-com frenzy at the turn of the millennium. Later, at Reuters, I ran an innovation group where our teams built wacky, sometimes brilliant products for hedge fund managers and Indian farmers alike. For several years, I worked with venture capitalists in Europe, backing the most ambitious technology founders we could find – and I still invest actively in young technology companies. As a start-up investor, I have spoken with hundreds of technology founders in fields as diverse as artificial intelligence, advanced biology, sustainability, quantum computing, electric vehicles and space flight.

But my academic training is in the social sciences. At university I focused on politics, philosophy and economics – though, unusually, I also took a programming course with a group of physicists who were much smarter than me. And for much of my career, my focus has been on how technology is transforming business and society. In my career as a journalist, first at *The Guardian* and then *The Economist*, I found myself having to explain complicated topics from the world of software engineering to a mainstream audience. And I've taken a particular interest in the political implications of new forms of technology. For a time, I was a non-executive member of Ofcom, the regulator that looks at the telecom, internet and media industries in the UK. In 2018, I became a board member of the Ada Lovelace Institute, where we have been looking at the ethical implications of the use of data and artificial intelligence in society.

Over the last few years, I have been channelling my attempts to straddle the 'two cultures' into *Exponential View* – a newsletter and podcast that explores the impact of new technology on society. I founded it after my third start-up, PeerIndex, was acquired by a much larger technology firm. PeerIndex applied machine learning techniques (on which, more later) to large amounts of public data about what people do online. We grappled with many ethical dilemmas about what it was and wasn't appropriate to do with this data. After my company's acquisition, I had the mental space to explore such issues in my newsletter.

Exponential View has resonated with people. At the time of writing, it has a readership of nearly 200,000 subscribers around the globe, ranging from some of the world's most well-known founders through to investors, policymakers and academics in more than 100 countries. And it has allowed me to delve into the most thought-provoking questions raised by new technology. Through my podcast series of the same name, I've conducted more than 100 interviews with engineers, entrepreneurs, policymakers, historians, scientists and corporate executives. Over more than six years, I've read tens of thousands of books, newspapers and magazine pieces, blog posts and journal articles as part of my research. I recently estimated that I have read more than 20 million words in the last half-decade in my effort to understand what is going on. (Fortunately, this book is somewhat shorter.)

The conclusion all this research has led me to is deceptively simple. At heart, the argument of *Exponential* has two key strands. First, new technologies are being invented and scaled at an ever-faster pace, all while decreasing rapidly in price. If we were to plot the rise of these technologies on a graph, they would follow a curved, exponential line.

Second, our institutions – from our political norms, to our systems of economic organisation, to the ways we forge relationships – are changing more slowly. If we plotted the adaptation of these institutions on a graph, they would follow a straight, incremental line.

The result is what I call the 'exponential gap'. The chasm between new forms of technology – along with the fresh approaches to business, work, politics and civil society they bring about – and the corporations, employees, politics and wider social norms that get left behind.

Of course, this only raises more questions. What effects do exponential technologies have in different spheres – from work, to conflict, to politics? For how long can this exponential change continue – will it ever stop? And what can we all do, as policymakers, business leaders or citizens, to prevent the exponential gap eroding our societies?

The structure of this book tries to make my answers as clear as possible. In the first part, I will explain what exponential technologies are and why they have come about. I argue that our age is defined by the emergence of several new 'general purpose technologies', each improving at an exponential rate. It's a story that starts with computing – but also encompasses energy, biology and manufacturing. The breadth of this change means that we have entered a wholly new era of human society and economic organisation – what I call the 'Exponential Age'.

Next, I move on to the implications this has for human society more broadly – the emergence of the exponential gap. There are many reasons why human-built institutions are slow to adapt, from the psychological trouble we have conceptualising exponential change, through to the inherent difficulty of turning around a big organisation. All contribute to the widening gulf between technology and our social institutions.

But what effects does the exponential gap have in practice? And what can we do about it? Those questions are the focus of the rest of this book. I'll take you from the economy and work, through the geopolitics of trade and conflict, to the broader relationship between citizens and society.

First, we'll explore what exponential technologies do to businesses. During the Exponential Age, technology-driven companies tend to become bigger than was previously thought possible – and traditional companies get left behind. This leads to winner-takes-all markets, in which a few 'superstar' companies dominate – with their rivals spiralling into inconsequentiality. An exponential gap emerges – between our existing rules around market power, monopoly, competition and tax, and the newly enormous companies that dominate markets.

I'll also show how the prospects of employees are changing thanks to the emergence of these companies. The relationships between workers and employers are always in flux, but now they are shifting more rapidly than ever. The superstar companies favour new styles of work, mediated by gig platforms, which may be problematic for workers. Existing laws and employment practices struggle to cope with the changing norms surrounding labour.

Second, we'll explore the transformation of geopolitics – discussing how exponential technologies are rewiring trade, conflict and the global balance of power. Here, two great shifts are underway. The first is a return to the local. New innovations alter the way we access commodities, manufacture products and generate energy – increasingly, we will be able to produce all three within our own regions. At the same time, the increasing complexity of our economies will make cities more important than ever, creating tension between regional and national governments. If the story of the industrial age was one of globalisation, the story of the Exponential Age will be one of re-localisation. The second is the transformation of warfare. As the world gets re-localised, patterns of global conflict will shift. Nations and other actors will be able to make use of new adversarial tactics, from cyber threats to drones and disinformation. These will dramatically reduce the cost of initiating conflict, making it much more common. A gap will emerge between new, high-tech forms of attack and societies' ability to defend themselves.

Third, we'll examine how the Exponential Age is rewiring the relationship between citizen and society. State-sized companies are on the rise – and they are challenging our most basic assumptions about the role of private corporations. Markets are metastasising across ever-greater swathes of the public sphere and our private lives. Our national conversations are increasingly conducted on privately owned platforms; intimate details about our innermost selves are bought and sold online, thanks to the emergence of the data economy; and even the way we meet friends and form communities has been turned into a commodity. But because we remain wedded to an industrial-age conception of the role of markets, we don't yet have the toolkit to prevent these changes eroding our most cherished values.

In other words, an exponential gap is challenging many elements of our society. But that's something we can address. And so, at the end of the book, I'll explain the broad principles that we need to make sure we thrive in an age of exponential change – from making our institutions more resilient to rapid transformation, to reiterating the power of collective ownership and decision-making. The resulting book is, I hope, a holistic guide to how technology is changing our society – and what we should do about it.

As I wrote this book, the world changed dramatically. When I first started my research, there was no such thing as Covid-19, and lockdowns were the remit of zombie apocalypse movies. But as I was halfway through writing my first draft, countries around the world began shutting their borders and issuing stay-at-home orders to their populations – all to prevent a virus wreaking havoc on their health systems and economies.

On one level, the pandemic felt distinctly low-tech. Lockdowns have been used for millennia to prevent the spread of disease. Quarantines are nothing new: the word derives from the time of the Black Death, when sailors had to isolate for 40 days before coming on shore. That the global economy was brought low by a virus reminded us of how many ancient problems technology hasn't yet been able to solve.

But the pandemic also hammered home some of the key points of this book. The spread of the virus demonstrated that exponential growth is hard to control. It creeps up on you and then explodes – one moment everything seems fine, the next your health service is on the verge of being overwhelmed by a new disease. And humans struggle to conceptualise the speed of that shift, as shown by the lackadaisical responses of many governments to the spread of coronavirus, particularly in Europe and America.

At the same time, the pandemic revealed the full power of recent inventions. In most of the developed world, lockdowns were only possible due to widespread access to fast internet. Those of us locked at home spent much of the pandemic glued to our phones. And, most strikingly of all, within a year scientists had developed dozens of new vaccines – which, as we'll see, were made possible by new innovations like machine learning. In some ways, exponential technology proved its mettle with Covid-19.

Above all, the pandemic revealed that Exponential Age technologies – whether video calls or social media platforms – are now embedded into every part of our lives. And this will only become more pronounced. As the rate of change speeds up, the interaction between technology and other spheres of our lives – from demography to statecraft to economic policy – will become increasingly constant. Neat distinctions between the realm of technology and the realm of, say, politics will become unhelpful. Technology is remaking politics, and politics is shaping technology. Any constructive analysis of either requires an analysis of both. And for politics, one could substitute economics, or culture, or business strategy.

As a result of the constant feedback loop of technology, economics, politics and society, making stable predictions about the future is difficult. Even as I wrote this book, its subject matter was constantly shifting – no sooner would I finish a chapter than it would need to be updated to incorporate new developments. Such are the perils of writing in an age of exponential change.

But my hope is that this book remains a useful introduction to where new technology is taking us. We are living in an era when technology is getting better, faster and more varied at a greater speed than ever before. That process is undermining the stability of many of the norms and institutions that define our lives. And we don't, at the moment, have a road map that will help us get to the future we want.

This book, by itself, is unlikely to offer a perfect map. But it might help reveal the terrain, and point us in the right direction.

Azeem Azhar
London, April 2021

1

The Harbinger

Before I knew what Silicon Valley was, I had seen a computer. It was December 1979, and our next-door neighbour had brought home a build-it-yourself computer kit. I remember him assembling the device on his living room floor and plugging it into a black-and-white television set. After my neighbour meticulously punched in a series of commands, the screen transformed into a tapestry of blocky pixels.

I took the machine in with all the wonder of a seven-year-old. Until then, I had only seen computers depicted in TV shows and movies. Here was one I could touch. But it was more remarkable, I think now, that such a contraption had even got to a small suburb of Lusaka in Zambia in the 1970s. The global supply chain was primordial, and remote shopping all but non-existent – and yet the first signs of the digital revolution were already visible.

The build-it-yourself kit piqued my interest. Two years later, I got my own first computer: a Sinclair ZX81, picked up in the autumn of 1981, a year after moving to a small town in the hinterlands beyond London. The ZX81 still sits on my bookshelf at home. It has the footprint of a 7-inch record sleeve and is about as deep as your index and middle fingers. Compared to the other electronic items in early-1980s living rooms – the vacuum-tubed television or large cassette deck – the ZX81 was compact and light. Pick-up-with-your-thumb-and-forefinger light. The built-in keyboard, unforgiving and taut when pressed, wasn't something you could type quickly on. It only responded to stiff, punctuated jabs of the kind you might use to admonish a friend. But you could get a lot out of this little box. I remember programming simple calculations, drawing basic shapes and playing primitive games on it.

This device, advertised in daily newspapers across the UK, was a breakthrough. For £69, we got a fully functional computer. Its simple programming language was, in principle, capable of solving any computer problem, however complicated (although it might have taken a long time).[1] But the ZX81 wasn't around for long. Technology was developing quickly. Within a few years, my computer – with its blocky black-and-white graphics, clumsy keyboard and slow processing – was approaching obsolescence. Within six years, my family had upgraded to a more modern device, made by Britain's Acorn Computers. The Acorn BBC Master was an impressive beast, with a full-sized keyboard and a numeric keypad. Its row of orange special-function keys wouldn't have looked out of place on a prop in a 1980s space opera.

If the exterior looked different to the ZX81's, the interior had undergone a complete transformation. The BBC Master ran several times faster. It had 128 times as much memory. It could muster as many as 16 different colours, although it was limited to displaying eight at a time. Its tiny speaker could emit up to four distinct tones, just enough for simple renditions of music – I recall it beeping its way through Bach's Toccata and Fugue in D Minor. The BBC Master's relative sophistication allowed for powerful applications, including spreadsheets (which I never used) and games (which I did).

Another six years later, in the early 1990s, I upgraded again. By then, the computer industry had been through a period of brutal consolidation. Devices like the TRS-80, Amiga 500, Atari ST, Osborne 1 and Sharp MZ-80 had vied for success in the market. Some small companies had short-lived success but found themselves losing out to a handful of ascendant new tech firms.

It was Microsoft and Intel that emerged from the evolutionary death match of the 1980s as the fittest of their respective species: the operating system and the central processing unit. They spent the next couple of decades in a symbiotic relationship, with Intel delivering more computational power and Microsoft using that power to deliver better software. Each generation of software taxed the computers a little more, forcing Intel to improve its subsequent processor. 'What Andy giveth, Bill taketh away' went the industry joke (Andy Grove was Intel's CEO; Bill Gates, Microsoft's founder).

At the age of 19 I was oblivious to these industry dynamics. All I knew was that computers were getting faster and better, and I wanted to get hold of one. Students tended to buy so-called PC clones – cheap, half-branded boxes which copied the eponymous IBM Personal Computer. These were computers based on various components that adhered to the PC standard, meaning they were equipped with Microsoft's latest operating system – the software that allowed users (and programmers) to control the hardware.

My clone, an ugly cuboid, sported the latest Intel processor, an 80486. This processor could crunch through 11 million instructions per second, probably 4–5 times more than my previous computer. A button on the case marked 'Turbo' could force the processor to run some 20 per cent faster. Like a car where the driver keeps their foot on the accelerator, however, the added speed came at the cost of frequent crashes.

This computer came with 4 megabytes of memory (or RAM), a 4,000-fold improvement on the ZX81. The graphics were jaw-dropping, though not state-of-the-art. I could throw 32,768 colours on the screen, using a not-quite cutting-edge graphics adaptor that I plugged into the machine. This rainbow palette was impressive but not lifelike – blues in particular displayed poorly. If my budget had stretched £50 more, I might have bought a graphics card that painted 16 million colours, so many that the human eye could barely discern between some of the hues.

The 10-year journey from the ZX81 to PC clone reflected a period of exponential technological change. The PC clone's processor was thousands of times more powerful than the ZX81's, and the computer of 1991 was millions of times more capable than that of 1981. That transformation was a result of swift progress in the nascent computing industry, which approximately translated to a doubling of the speed of computers every couple of years.

To understand this transformation, we need to examine how computers work. Writing in the nineteenth century, the English mathematician and philosopher George Boole set out to represent logic as a series of binaries. These binary digits – known as 'bits' – can be represented by anything, really. You could represent them mechanically by the positions of a lever, one up and one down. You could, theoretically, represent bits with M&Ms – some blues, some reds. (This is certainly tasty, but not practical.) Scientists eventually settled on 1 and 0 as the best binary to use.

In the earliest days of computing, getting a machine to execute Boolean logic was difficult and cumbersome. And so a computer – basically any device that could conduct operations using Boolean logic – required dozens of clumsy mechanical parts. But a key breakthrough came in 1938, when Claude Shannon, then a master's student at the Massachusetts Institute of Technology, realised electronic circuits could be built to utilise Boolean logic – with on and off representing 1 and 0. It was a transformative discovery, which paved the way for computers built using electronic components. The first programmable, electronic, digital computer would famously be used by a team of Allied codebreakers, including Alan Turing, during World War Two.

Two years after the end of the war, scientists at Bell Labs developed the transistor – a type of semiconductor, a material that partly conducts electricity and partly doesn't. You could build useful switches out of semiconductors. These in turn could be used to build 'logic gates' – devices that could do elementary logic calculations. Many of these logic gates could be stacked together to form a useful computing device.

This may sound technical, but the implications were simple: the new transistors were smaller and more reliable than the valves that were used in the earliest electronic components, and they paved the way for more sophisticated computers. In December 1947, when scientists built the first transistor, it was clunky and patched together with a number of large components, including a paper clip. But it worked. Over the years, transistors would become less ad hoc, and more consistently engineered.

From the 1940s onwards, the goal became to make transistors smaller. In 1960, Robert Noyce at Fairchild Semiconductor developed the world's first 'integrated circuit', which combined several transistors into a single component. These transistors were tiny and could not be handled individually by man or machine. They were made through an elaborate process a little like chemical photography, called photolithography. Engineers would shine ultraviolet light through a film with a circuit design on it, much like a child's stencil. This imprints a circuit onto a silicon wafer, and the process can be repeated several times on a single wafer – until you have several transistors on top of one another. Each wafer may contain several identical copies of circuits, laid out in a grid. Slice off one copy and you have a silicon 'chip'.

One of the first people to understand the power of this technology was Gordon Moore, a researcher working for Noyce. Five years after his boss's invention, Moore realised that the physical area of integrated circuits was reducing by about 50 per cent every year, without any decrease in the number of transistors. The films – or 'masks' – used in photolithography were getting more detailed; the transistors and connections smaller; the components themselves more intricate. And this reduced costs and improved performance. Newer chips, with their smaller components and tighter packing, were faster than older ones.

Moore looked at these advances and in 1965 he came up with a hypothesis. He postulated that these developments would double the effective speed of a chip for the same cost over a certain period of time.[2] He eventually settled on the estimate that, every 18–24 months, chips would get twice as powerful for the same cost. Moore went on to co-found Intel, the greatest chip manufacturer of the twentieth century. But he is probably more famous for his theory, which became known as 'Moore's Law'.

This 'law' is easy to misunderstand; it is not like a law of physics. Laws of physics, based on robust observation, have a predictive quality.

Newton's Laws of Motion cannot be refuted by everyday human behaviour. Newton told us that force equals mass times acceleration – and this is almost always true.[3] It doesn't matter what you do or don't do, what time of day it is, or whether you have a profit target to hit.

Moore's Law, on the other hand, is not predictive; it is descriptive. Once Moore outlined his law, the computer industry – from chipmakers to the myriad suppliers who supported them – came to see it as an objective. And so it became a 'social fact': not something inherent to the technology itself, but something wished into existence by the computer industry. The materials firms, the electronic designers, the laser manufacturers – they all wanted Moore's Law to hold true. And so it did.[4]

But that did not make Moore's Law any less powerful. It has been a pretty good guide to computers' progress since Moore first articulated it. Chips did get more transistors. And they followed an exponential curve: at first getting imperceptibly faster, and then racing away at rates it is hard to comprehend.

Take the below graphs. The top one shows the growth of transistors per microchip from 1971 to 2017. That this graph looks moribund until 2005 reflects the power of exponential growth. On the second graph, which shows the same data using a logarithmic scale – a metric that converts an exponential increase into a straight line – we see that,

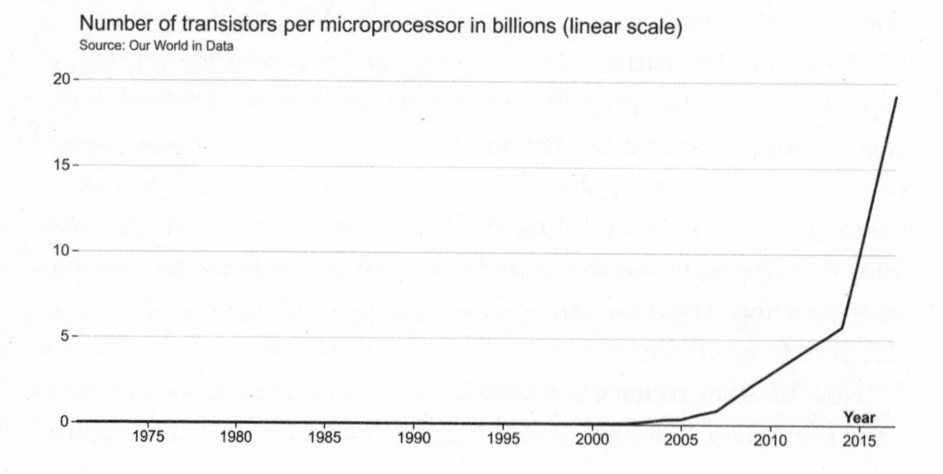

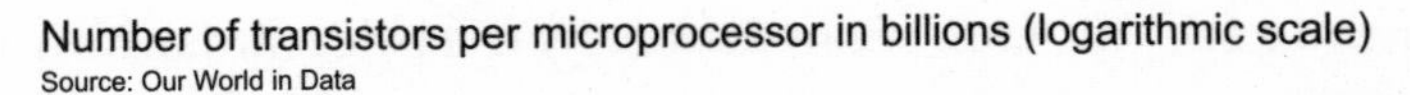

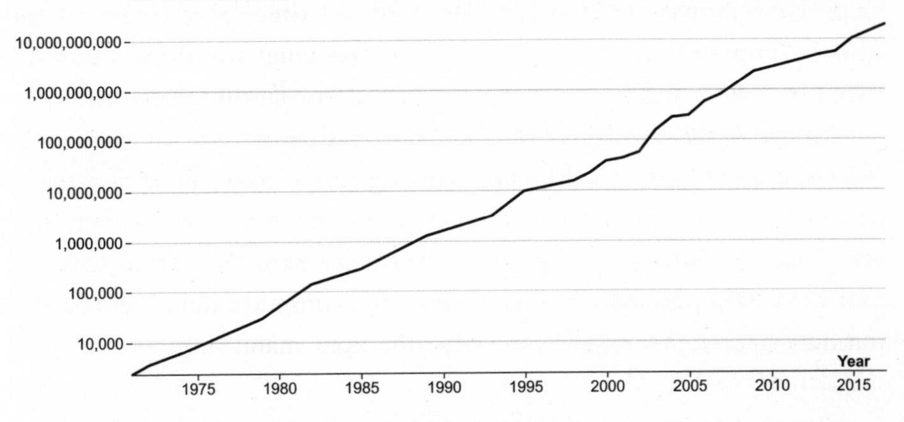

between 1971 and 2015, the number of transistors per chip multiplied nearly 10 million times.

The magnitude of this shift is almost impossible to conceptualise, but we can try to grasp it by focusing on the price of a single transistor. In 1958, Fairchild Semiconductor sold 100 transistors to IBM for $150 apiece.[5] By the 1960s, the price had fallen to $8 or so per transistor. By 1972, the year of my birth, the average cost of a transistor had fallen to 15 cents,[6] and the semiconductor industry was churning out between 100 billion and 1 trillion transistors a year. By 2014, humanity produced 250 billion billion transistors annually: 25 times the number of stars in the Milky Way. Each second, the world's 'fabs' – the specialised factories that turn out transistors – spewed out 8 trillion transistors.[7] The cost of a transistor had dropped to a few billionths of a dollar.

Why does this matter? Because it led computers to improve at an astonishing rate. The speed that a computer can process information is roughly proportional to the number of transistors that make up its processing unit. As chips gained transistors, they got faster. Much faster. At the same time, the chips themselves were getting cheaper.

This extraordinary dropping in price is what drove the computing revolution of my teenage years, making my BBC Master so much better than my ZX81. And since then, it has transformed all of our lives again.

When you pick up your smartphone, you hold a device with several chips and billions of transistors. Computers – once limited to the realms of the military or scientific research – have become quotidian. Think of the first electronic computer, executing Alan Turing's codebreaking algorithms in Bletchley Park in 1945. A decade later, there were still only 264 computers in the world, many costing tens of thousands of dollars a month to rent.[8] Six decades on, there are more than 5 billion computers in use – including smartphones, the supercomputers in our pockets. Our kitchen cupboards, storage boxes and attics are littered with computing devices – at only a few years old, already too outdated for any modern application.

Moore's Law is the most famous distillation of the exponential development of digital technology. Over the last half-century, computers have got inexorably faster – bringing with them untold technological, economic and social transformations. The goal of this chapter is to explain how this shift has come about, and why it looks set to continue for the foreseeable future. It will also serve as an introduction to the defining force of our age: the rise of exponential technologies.

Put simply, exponential growth refers to an increase that compounds consistently over time. Whereas a linear process is what happens to your age, increasing by a predictable one with each revolution of the earth around the sun, an exponential process is like a savings account with interest. It increases by a fixed percentage, say 2 per cent every year. But next year's 2 per cent applies not just to your original savings but to your savings plus last year's interest. Such compounding starts slow – it's even slightly boring. But at some point, the curve turns upwards and starts to take off. From this point onwards, the value leaps higher at a dizzying pace.

Any number of natural processes follow an exponential pattern: the number of bacteria in a petri dish, or the spread of a virus through a population, for example. A more recent development, however, is the emergence of exponential technologies. I define an exponential technology as one that can, for a roughly fixed cost, improve at a rate of more than 10 per cent per year for several decades. A mathematical purist, of course, would argue that even a 1 per cent compounding change is an exponential change. Strictly speaking, it is. But a 1 per cent annual change takes a lot of time to get going. A figure compounding at 1 per cent per annum would take 70 years, the better part of a lifetime, to double.

That is why the threshold of 10 per cent per year is important. A 10 per cent compounding improvement in the price and performance of a technology would result in it becoming more than 2.5 times more powerful for the same price every 10 years. Conversely, the cost would drop by more than three-fifths for the same level of performance. A decade is only two traditional business-planning cycles, well within the span of a single job or a part of a career. It is the lifetime of two parliaments in the UK or France, three in Australia, or two and a bit presidential terms in the US.

The second part of my definition is also crucial. For a technology to be exponential, this change should hold true for decades – and not just be a short-lived trend. A technology that advances at more than 10 per cent for a few years, and then stops, would be much less transformative than one that constantly develops. For this reason, the diesel engine is not an exponential technology. In their early years, diesel engines improved rapidly. But the improvements soon petered out. On the other hand, the computer chip business – with its roughly 50 per cent annual improvement over five decades – certainly qualifies.

Picture yourself changing your car after ten years of ownership. And imagine if the main features of the car – its top speed or fuel efficiency, say – had improved at 10 per cent per annum. Your new wheels might have double the fuel efficiency or more than double the top speed. Generally, that doesn't happen. But for many of the technologies discussed in this book, it is exactly what happens. In fact, a number of these technologies effectively improve at rates of 20 to 50 per cent (or more) per year. An innovation rate like this means that, over a decade, you would experience between a sixfold and a 60-fold increase in capability for the same price.

There are two sides to this phenomenon: decreases in price, and increases in potential. As the price of a technology drops, it starts to crop up everywhere. Industry can suddenly afford to bundle exponential technologies into new products. Humans first put chips into specialist devices bought by the military and space agencies, and then into minicomputers only affordable for the largest companies. Desktop computers followed a decade later, and as chips got cheaper and smaller, they were put into phones.

At the same time, the power of the technology explodes. The capabilities of a typical smartphone – high-definition colour video, high-fidelity sound, fast-action video games, scanners that transcribe text – were not available to anyone, not even the richest countries, just a couple of decades ago. When technologies develop exponentially, they lead to continually cheaper products which are able to do genuinely new things.

To see this process in action, it's worth examining the work of Horace Dediu. A business analyst, Dediu trained with Clayton Christensen, the world-renowned Harvard academic who wrote the bible of many a Silicon Valley tech firm, *The Innovator's Dilemma*. Dediu carries a well-deserved reputation in his own right, thanks to his research into patterns of innovation. Over the last two decades, he has analysed more than 200 years of historical data, to examine how quickly technologies have spread through the American economy.[9] He casts a wide net, meticulously tracking a broad range of innovations – flush toilets, the electrification of printing, the spread of roads, the vacuum cleaner, diesel locomotives, power steering in cars, electric arc furnaces, artificial fibres, ATM machines, digital cameras, social media and computer tablets, among many others. For each, he has established how long it took to reach 75 per cent penetration of the American market, meaning three-quarters of adults (or households, if appropriate) have access to it.

While each product is different, there are common themes to the ways they take off. The spread of most technologies follows a 'logistic curve', or S-curve. At first, the uptake of a technology is slow. Early adopters are experimenting with it; while producers are figuring out exactly what to make and how to price it, and are building up their capacity. At some point the product hits an inflection point, and its rate of spread increases very rapidly. So the first two parts of the curve look like a classic exponential curve: slow and boring at first, then rapid and exciting. Unlike a pure exponential curve, however, the S-curve has a limit. After all, there are only so many cars or washing machines a family can own. As the market gets saturated, the uptake tapers off: there are fewer families who don't own a digital camera or microwave, or fewer steelmakers who haven't moved to electric arc furnaces. The steep part of the graph starts to flatten. In other words, the pattern of uptake looks like a lazy 'S'.

Sometimes, the point of market saturation ends up being more distant than almost anyone expected. In 1974, Bill Gates said he envisaged a 'computer on every desk and in every home'. At that time, there were fewer than 500,000 computers, of any type, worldwide. By the turn of the millennium, the number of computers exceeded 500 million – still fewer than one device per European or American home. Yet, within a couple of decades, a typical family in the West had half a dozen computers at home – between smartphones, the family computer, a modern TV, and a smart speaker like an Amazon Alexa. A gadget-savvy household could easily break into the double digits.

In general, the S-curve model remains accurate. When you're dealing with exponential technologies, however, the speed of their acceleration up the 'S' can be astonishing. The process of market saturation has been getting faster for decades, and this increasing rate of change will have been noticeable to any American who lived through the twentieth century. Someone born in 1920 would live for a little less than 55 years, probably witnessing the moon landing but possibly not Nixon's downfall. With a few exceptions – the atom bomb, space flight – the technology they encountered would have remained fairly constant: cars, phones, televisions, washing machines, electricity and flush toilets.[10] Some products invented relatively early, like the microwave – first sold in 1946 – remained vanishingly rare even by the 1970s.[11]

For someone born in the age of Moore's Law, the picture is different. Products roll out far faster, and the technologies powered by digital infrastructure are fastest of all. It took 11 years for social media to reach 7 in 10 Americans, at a time when the average lifespan of those alive when it was introduced exceeded 77 years. So social media took 14 per cent of a lifespan to achieve saturation. The comparative metric for electricity was 62 per cent of the average lifespan. Benchmarked against average lifespan at time of introduction, smartphones diffused 12.5 times faster than the original telephone.

The chart opposite makes the point even more simply. To the right are the defining technologies of the early twentieth century: the telephone, electric power and the automobile. Each was introduced at the turn of the century and took over 30 years to reach three-quarters of American households – at a time when average life expectancy was around 50. To the left are a handful of the first exponential technologies of our time, all undergirded by growing computing power. Each found their way into three-quarters of American homes within 8–15 years, at a time when life expectancy at birth exceeded 75 years. And it is not just that technologies of the present scale faster than those of the past: the pace with which they scale is constantly increasing.

In other words, technology – and particularly digital technology – diffuses at a faster rate than ever before. And this process is continually getting quicker. Life in the age of Moore's Law is defined by the exponential spread of technology.

If this acceleration has been underway for half a century, only in the last decade or two has it been obvious. Take social media. The first web-based social network in the world was SixDegrees. I joined a couple of days after its launch in 1997. Friendster and LinkedIn, where I was among the first 1,000 members, both appeared in 2003; MySpace

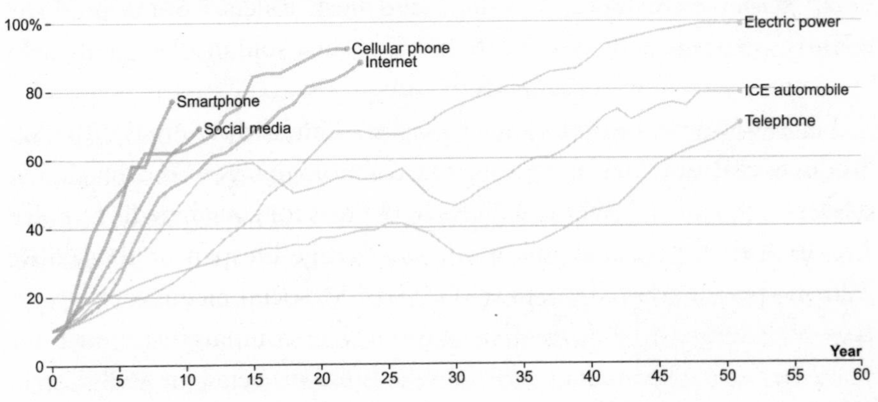

later that year. MySpace grew quickly and became the titan of this emerging industry, having, at its peak, 115 million users. But it was Facebook that demonstrated the sheer speed with which digital technologies can grow. After Facebook launched in February 2004, it became one of the fastest-growing products in world history, amassing its first million users in just 15 months. Mark Zuckerberg, its founder, today has the most popular product on the planet. Facebook had 2.5 billion users at the end of 2019.

Today, however, the speed of Facebook's growth looks positively quaint. Witness Lime. Founded in San Francisco in January 2017, the company places distinctive green electric scooters and bikes around cities. Press a button on your smartphone and you can hire one by the minute. Although Lime's business is much more complicated than Facebook's – incorporating bikes that need GPS, GSM connections, charging, maintenance and tracking – it took a mere six months for Lime to serve 1 million rides; and only another seven months to get to 10 million.[12] All this was made possible by the revolution in computing: prices had fallen so much, Lime could put a small computer and GSM radio into every one of the hundreds of thousands of bikes they operate.[13]

The increasingly rapid growth of digital technologies is not limited to the US. KakaoTalk is the leading Korean social network, the equivalent to WeChat or WhatsApp. In January 2016, the firm decided to launch a bank. Within two weeks, 2 million Koreans – about 4 per cent of the country – had opened an account. By the summer of 2019, more than 20 per cent of Koreans had done so.[14] And as soon as we get to grips with one fast-moving exponential age product, another shows up. Take TikTok, a social network for funny videos. It went from an unheard-of service to the most downloaded app in the world in a matter of months. And with that growth came an unparalleled flow of sales. ByteDance, Tiktok's parent company, reported sales of $7 billion in 2018; two years later its revenues had more than quintupled. For comparison, just five years earlier, Facebook had exceeded the same milestone of $7 billion in revenue; in its next two years, its revenues had only tripled.[15]

This increasing speed is the legacy of Moore's Law. The hardware that underpins digital technology lends itself to continual increases in power and continual decreases in price. As chips develop at exponential rates – 50 per cent or more, compounding every year for many years – they give access to unimaginable computing power, for trivial sums of money. That hyperdeflation creates ever-greater possibilities: new products that can, in turn, spread faster through our economies. The overall process is one of constant acceleration.

By the early years of the twenty-first century, some technologists had started to notice the slowdown of Moore's Law. That was perhaps unsurprising. Technologies don't tend to improve exponentially indefinitely. Cars aren't much faster today than they were at the end of World War Two. Modern passenger aircraft trundle along at 500 or so miles per hour – not much quicker than the 468 miles per hour the first passenger jets managed in the 1950s.

And there's a strong case that our current approach to chip design is close to bumping against the limits of what is possible. Scientists have come up with increasingly complicated processes to meet Moore's predictions. As the transistors get smaller, they require ever-more precise machines to build them: today's semiconductor factories depend on extraordinarily sophisticated laser technology, with the most advanced lasers used costing more than $100 million apiece. At the same time, any tiny change to the factories' atmospheric conditions has come to pose an existential threat to their microscopic transistors: a single speck of dust can ruin the silicon wafers. And so the rooms in which chips are made are today the most placid in the world, sitting on legions of anti-vibration dampers. They are also the cleanest. The air in these rooms, which sometimes approach 200,000 square feet in area, is often filtered some 600 times an hour. (By comparison, a hospital operating theatre only needs its air cleaned 15 times an hour.)

This is what is meant when we say that Moore's Law is a social fact rather than a hard law: the semiconductor industry has been hell-bent on meeting it. Some economists estimate that the amount of research required to keep Moore's Law ticking over increased 18-fold between 1971 and 2018. The construction costs of semiconductor fabs have increased at about 13 per cent a year – the most recent cost $15 billion or more to build.[16]

But, despite the industry's efforts, by the late 2010s the growth of transistors per chip was beginning to slow. Like sweaty commuters on a hot day, cheek to armpit, these submicroscopic circuits were irritating each other. Each miniscule transistor generates heat that can spill over to neighbouring circuits and make them unreliable – a problem that chip engineers are finding increasingly hard to counteract. What's more, modern transistors are so small – only a few atoms wide – that they may soon be prone to the spooky behaviour of quantum physics. At this scale, particles are so small that they behave like waves, meaning they can pass through physical barriers and stumble into places they shouldn't be. Moore's Law is being derailed by quantum-drunk electrons.

But this doesn't mean that the growth of computing power will decelerate. The computing revolution shows no sign of slowing down. Ray

Kurzweil, one of the world's leading technology researchers, posits a theory of technological development that seeks to explain why. Technology, he says, tends to develop at an accelerating pace – according to what he calls 'the Law of Accelerating Returns'. At the heart of Kurzweil's model is a positive feedback loop. Good computer chips allow us to crunch more data, which help us to learn how to make better computer chips. We can then use these new chips to help us build better chips, and so on. In Kurzweil's view, this process is constantly accelerating: the returns of each generation of technology layer on top of the previous generation's, and even feed into one another.[17]

However, the crucial part of Kurzweil's theory isn't about any one technology, like the automobile or the microchip. His focus is on how different technologies interact. Kurzweil's great insight is that the exponential march of technology is not, in fact, about the straightforward progression of individual inventions, or even individual sectors of the economy. In fact, the illusion of continual exponential technological development is down to dozens of different technologies developing in tandem and continually interacting.

Remember the S-curves in Horace Dediu's data. When a technology is first created, its development and spread follows a shallow gradient. That speaks of slow but meaningful progress. At some point, however, the development of the technology picks up pace. Rapid expansion follows, until, at some stage, progress peters out. Our once near-vertical graph flattens back to the horizontal.

In Kurzweil's view, however, at any one time there are multiple technologies following an S-curve. As one S-curve reaches its steepest gradient, another curve begins. Once our first curve starts to flatten out, the younger technology is approaching the explosive phase of its acceleration – and takes up the mantle of rapid growth. And, most importantly of all, these different technologies nourish one another – innovations in one sector inspire developments in the next. When one technology reaches the limits of its potential, a new technological paradigm is waiting in the wings to pick up the slack. The result is that, across a society, there is a quickening in the pace of technological progression – even if the development of individual technologies is consistently slowing off.[18]

This theory has profound implications for the future of computing. While the paradigm described by Moore's Law is reaching its limits, we are not reaching the limits of computational power in general. We should always find some new approach that can help meet the growing demands of users, the theory suggests. It's just that, in the future, increased computing power might not be based on cramming more transistors onto a chip.

So far, Kurzweil's theory seems to be holding up. In the first years of the new millennium, at around the same time we approached what many engineers thought were the physical limits of Moore's Law, we reached a tipping point. Finally, there was enough data and enough computational power to allow for the development of a new technical paradigm: artificial intelligence. And it catalysed a wholly new way of thinking about computing power – one that has broken through the limits of our earlier method of chip design.

Humans have pondered the possibility of building artificial intelligence since antiquity. According to Stuart Russell, the author of the top graduate textbook on AI, a computer can be considered intelligent if it is able to take actions that can achieve its objectives.[19] Crucially, a piece of AI software needs to be able to make some kind of decision – rather than just blindly following each step in its program code.

After the term 'artificial intelligence' was coined in 1955 by a computer scientist, John McCarthy, researchers set out to build just such 'intelligent' machines. Over the next 60 years, AI research progressed slowly. The field had many false starts – with seemingly important breakthroughs fostering overinflated expectations, which in turn led to failure and despondency. The problem was a lack of data and a lack of computing power. For decades, many scientists believed that any major breakthrough in AI was probably going to come through an approach called 'machine learning'. This is a method that involves gathering huge amounts of information about a problem, and using algorithms to identify recurrent patterns. For example, you might teach an AI system the difference between a cat and a dog by showing it 10 million photos of cats and dogs, and telling the machine explicitly which are cats and which are dogs. Eventually, the 'model' will learn to discriminate between a picture of a cat and a picture of a dog. But until relatively recently, we lacked the data and computational power to realise the potential of such machine learning. Thanks to the amount of number-crunching involved, this approach needs masses of information and expensive computation. And that information and computing power didn't exist.

By the early 2010s, however, that had begun to change. There was suddenly a cornucopia of data, created by ordinary people sharing photos of their lives on the internet. At first this data wasn't particularly useful for AI researchers – until a professor called Fei-Fei Li set out to change that. Based at Stanford University, Li is a computer scientist specialising in the intersection of neuroscience and computer science – with a particular interest in how humans perceive objects. In 2009, driven by the idea that digitally mapping out as many real-world objects as possible would improve AI, Li set up ImageNet – a project that over a period of five years would single-handedly catalyse the explosion of useful AI. The site took the form of a meticulously detailed collection of 14,197,122 images, all hand-annotated with tags like 'vegetable', 'musical instrument', 'sport' and, yes, 'dog' and 'cat'. This dataset was used as the basis for an annual competition to find the algorithm that could most consistently and accurately identify objects. Thanks to ImageNet, good-quality labelled data was suddenly in high supply.

Alongside the profusion of data came an explosion in computing power. By 2010, Moore's Law had resulted in enough power to facilitate a new kind of machine learning, 'deep learning', which involved creating layers of artificial neurons modelled on the cells that underpin human brains. These 'neural networks' had long been heralded as the next big thing in AI. Yet they had been stymied by a lack of computational power. Not any more, however. In 2012, a group of leading AI researchers – Alex Krizhevsky, Ilya Sutskever and Geoffrey Hinton – developed a 'deep convolutional neural network' which applied deep learning to the kinds of image-sorting tasks that AIs had long struggled with. It was rooted in extraordinary computing clout. The neural network contained 650,000 neurons and 60 million 'parameters', settings you could use to tune the system. It was a game-changer. Before AlexNet, as Krizhevsky's team's invention was called, most AIs that took on the ImageNet competition had stumbled, for years never scoring higher than 74 per cent. AlexNet had a success rate as high as 87 per cent. Deep learning worked.

The triumph of deep learning sparked an AI feeding frenzy. Scientists rushed to build artificial intelligence systems, applying deep neural networks and their derivatives to a vast array of problems: from spotting manufacturing defects to translating between languages; from voice recognition to detecting credit card fraud; from discovering new medicines to recommending the next video we should watch. Investors opened their pocketbooks eagerly to back these inventors. In short order, deep learning was everywhere. As a result, neural networks demanded increasing amounts of data and processing power. A 2020 neural net, GPT-3 – used to generate text that could sometimes pass for being written by a human – had 175 billion parameters, about 3,000 times more than AlexNet.

But if the new approach to computing was AI, what was powering it? Between 2012 and 2018, the amount of computer power used to train the largest AI models increased about six times faster than the rate of Moore's Law. The graph over the page plots the growth in computation used for state-of-the-art AI systems, set against the exponential curve of Moore's Law. If AI's usage of computational power had followed the Moore's Law curve, it would have increased about seven times in six years. In fact, usage increased 300,000 times.[20]

It's an astonishing statistic. But it can be explained by precisely the process that Ray Kurzweil identified decades earlier. At just the moment we were bumping up against the limits of our old method – cramming more transistors onto a chip – scientists came up with a novel solution, drawing on a slightly different approach.

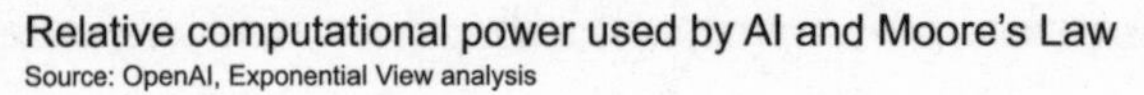

Relative computational power used by AI and Moore's Law
Source: OpenAI, Exponential View analysis

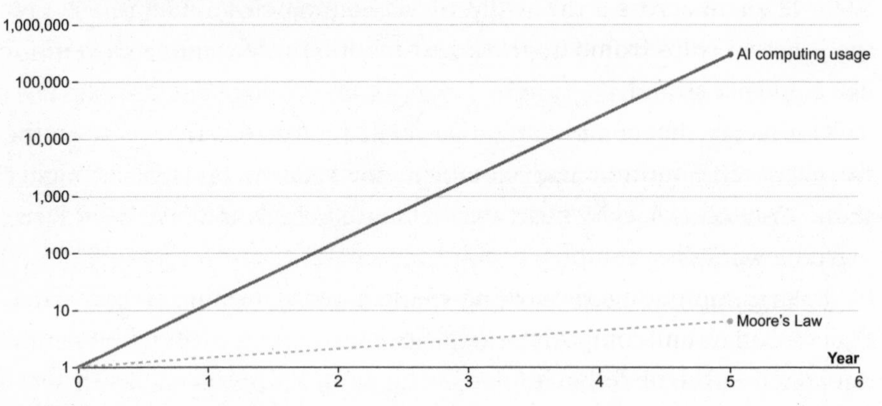

The answer lay in the kind of chips used. AI researchers like Krishevsky replaced traditional computer chips with ones that had been developed to draw high-end graphics for video games. You wouldn't use chips like these for everyday computing, but they proved surprisingly well suited for AI. In particular, they were remarkably good at maths. The calculations that video games needed to produce realistic scenes involved a lot of multiplying. To get a sophisticated neural net to work, you'd need to perform millions, sometimes billions, of such multiplications – and these graphics chips were up to the task.

As it became clear that there was an expanding market for these chips, the computer industry rose to the challenge. AI developers needed more power, and that power would come from these specialist chips. Firms like California's Cerebras and the UK's Graphcore began to build chips suited for this one task alone: running neural networks at high speeds.

The result has been the continued exponential growth of computing power, except free from the shackles of Moore's Law. That law is driven by miniaturisation – ever more transistors in an ever-smaller space. But today's AI chips don't rely on increasingly Lilliputian forms of engineering. In fact, some of these chips are made up of much bigger components. A traditional processor of the type you might find in your laptop has components some 7 nanometres apart – you could fit about 3,000 of them across a the width of single strand of human hair. The specialist AI chips from Graphcore are mapped at 16 nanometres– more like 1,300 per strand.

This means that computing power looks set to grow exponentially for the foreseeable future. And should our new forms of chip eventually prove unsuited to society's growing demands for computing power, waiting in the wings is a completely novel approach – 'quantum computing'.[21] In classical computing, the core unit of information is the bit, the binary digit. In quantum computing, that elementary unit is the qubit – and qubits reflect the underlying maths of quantum physics. Unlike the binary nature of bits, which must either be a 0 or a 1, a qubit can represent all values between 0 and 1 simultaneously.

Like specialist AI chips, quantum computers are not suitable for most types of computation, but they can tackle vital problems. For example, scientists hope quantum computers will allow us to create nitrogen fertilizers without releasing scads of carbon dioxide into the atmosphere. This is all about modelling new kinds of molecules, to be used as catalysts in the process of fertilizer production. Classical computers would take hundreds of thousands of years to model such molecules; a quantum computer would take about a day.[22]

In one experiment, reported in October 2019, a prototype quantum computer from Google performed a computing test in 200 seconds. The same test would have taken a state-of-the-art classical supercomputer around 10,000 years. This primordial quantum computer was more than 1 billion times faster than its classical rival.

In other words, so far Kurzweil has been vindicated. Even as one technological paradigm runs out of steam, new alternatives are being developed – so as we approach the bounds of one approach, another one takes its place. Whenever any single technology seems to be reaching a hard limit, inventive folk will draw on novel techniques – probably from adjacent sectors and disciplines – to overcome the problem.

And if this theory holds true in general, it feels particularly apt for computing. From the ZX81 to Google's prototype quantum computers, over the last half-century computers have been getting exponentially faster. This exponentiality shows no sign of slowing down: as computers speed up, new technological paradigms will develop. As one approach reaches its full potential, another will be waiting to pick up the slack.

In this sense, computing was the harbinger for the Exponential Age, the first technology to exhibit this remarkable rate of change. It showed us what happens when technologies accelerate, getting better and cheaper, creating entire industries in their wake.

But computers are not the only exponential technology. Across many sectors, technologies are improving at increasing speeds, and bringing massive changes with them. An eclectic mix of social, economic and political forces is driving the acceleration of these technologies. And it is this wider process of acceleration – and its causes – to which we now turn.

2

The Exponential Age

I know I am in the minority when I confess that *The Man with the Golden Gun* is one of my favourite Bond movies.[1] The two-dimensionality of the film's characters is rivalled only by that of its set design. But it has a certain charm. James Bond – played with characteristic comedy by Roger Moore – chases a freelance assassin, Francisco Scaramanga, across a series of beautiful tropical locations, from Macau to Thailand. Scaramanga has stolen the Solex Agitator, a device that can harness the sun's power to provide the world with electricity. Bond is tasked with getting it back. The whole movie is an exercise in techno-fetishism, symbolised by the futuristic solar power plant on Scaramanga's private island – where he and Bond eventually face off.

Released in 1974, the film reflected the anxieties of a world in the grips of an unprecedented oil crisis. In October 1973, a group of oil-exporting Middle Eastern countries had announced an embargo, refusing to export petroleum to the US and a number of its allies over their alleged support for Israel in the Yom Kippur War of that year. Over the next six months, oil prices rocketed to three times their October level. And against the backdrop of punishingly high fuel prices, alternative forms of energy suddenly captured the world's imagination. Hence the Solex Agitator, perhaps.

Yet despite Bond's sterling efforts murdering and romancing his way across the Asia-Pacific, solar power was not going to solve the oil crisis. Like the film's producers, the Japanese and American governments sprung into action, responding to rising oil prices and fuel shortages by doubling down on solar energy research. But, at the time, generating any useful amount of electricity from the sun's rays was simply too expensive.

In 1975, a silicon photovoltaic module – the shiny reflective substrate you now see adorning solar farms around the globe – cost about $100 per watt of power it could produce. At that price, solar power would remain an energy source for low-energy playthings and government gadgets, and little else. Satellites had been festooned with solar panels – which were lighter than chemical batteries – since 1958.[2] Some wristwatches and calculators sported them by the end of the 1970s. But for more widespread use, forget it. Solar power was prohibitively expensive.

That calculus has changed. Between 1975 and 2019, photovoltaics dropped in price some 500 times – to under 23 cents per watt of power.[3] The bulk of that change has come in the last decade. Even in 2010, it cost 30–40 cents (about 20–30 pence) to produce a kilowatt-hour of electricity using solar panels, making it 10–20 times more expensive than fossil fuels.[4] But the cost of solar power has been declining at exponential rates – about 40 per cent per annum for large commercial contracts. By October 2020, the cost of generating electricity from large-scale solar power, as well as from wind, had dropped below the cheapest form of fossil fuel product: combined-cycle gas generation.[5] In the decade up to 2019, the price of electricity generated by solar power had declined by 89 per cent.

No other form of electricity production can compete. From being entertainingly utopian in Scaramanga's time, and still fundamentally uneconomic at the start of the twenty-first century, solar power has become the cheapest source of electricity in two-thirds of the world.[6] Now, solar panels grace houses across the globe in rich nations and poor nations alike – in the north, in the south, in hot countries and cold. Cheap solar electricity has even transformed the lives of opium farmers in Afghanistan – powering irrigation, allowing for greater cultivation of the poppy, and leading to diversification into other crops.[7]

The story of solar power since the 1970s reveals that exponentiality is not limited to computing. Yes, the computer industry set the stage for a new era of accelerating technologies. But this phenomenon is not confined to transistors, nor to AI. Exponentiality has become widespread in four key domains of technology, which between them form the bedrock of the global economy. Computing, of course, is one. But so too are the domains of energy, biology and manufacturing. Each is undergoing a spectacular, thrilling transformation. The costs of the key technologies in each area are falling dramatically, by the equivalent of a factor of six or more every decade. We saw what exponentiality looks like in the first of these areas in the previous chapter. In this chapter, we'll explore the other three.

The breadth of these four areas demonstrates that we need to adopt a wide-lens view of exponential change. To make sense of exponentiality beyond the world of computing, we must view it as the result of a broad array of technological, economic and political forces. Explaining these forces is the goal of this chapter.

First, we'll examine why new technologies in each of these domains are having such a transformative impact across society. Second, we'll examine why they have come about. As we'll see, growing demand has pushed the price of technology ever lower. And every new innovation interacts – and combines – with the others to create new technological possibilities. All the while, growing networks of information and trade make the spread and scaling of new technology easier by the day.

Between these four key areas – computing, energy, biology and manufacturing – it is possible to make out the contours of a wholly new era of human society. We are not talking about a smattering of accelerating technologies in a few distinct areas. Exponentiality is baked into the very logic of technology, across numerous sectors of the economy. In other words, we are in the Exponential Age.

Let's look at the three remaining areas of exponential change one by one. First up, energy. Solar is not the only power source on an exponential trajectory. There's an economic reason that massive wind turbines increasingly occupy windy stretches of shallow water, gusty plains and mountaintops. Suddenly, wind power is cheap – and getting cheaper. In the 10 years to 2019, the cost of generating electricity from wind turbines declined by 70 per cent, or about 13 per cent per annum. This decrease is helping to displace other forms of energy production across the globe. Fossil fuel power stations are being mothballed, thanks to the exponential decline in the price of renewables.

Of course, wind and solar power alone cannot address all our energy needs. Their limitation lies in their intermittency – they generate power, but they do not offer an obvious way to store it. One of the remarkable features of fossil fuels is that they are stores of energy – a lump of coal is like a physical block of heat, to be transported wherever and burnt whenever its owner sees fit. The great challenge for renewables has been what to do at night, or on days without a strong breeze. But batteries are on an exponential trajectory too. The cost of lithium-ion storage dropped by 19 per cent per annum for the whole decade from 2010.[8] As of 2021, large-scale battery systems are nearly cheap enough to compete with coal- and gas-fired power stations. It is not unreasonable to expect that between 2020 and 2030, the price of electricity generated by solar power or wind power might decline several times more, and the price of battery storage at the same rate.

A similar process is at work in another of our key domains: biology. If anything, technological progress in the realm of biomatter makes the revolution in energy look glacial. For generations, biology seemed a difficult and messy science – one not particularly prone to technological transformation. If chemists have long been able to isolate simple chemicals in test tubes, and physicists to distil the universe down to its essential laws, biology has generally been fuzzier – focused as it is on the confusingly complex inner workings of living organisms.

Consider the process of reading the human genome, the genetic code that underpins each one of us. It is a complex business. The genome of a human is a long chain of binary information, some 3 billion letters long. We use the letters 'G', 'T', 'C' and 'A' to represent its chemical bases – guanine, thymine, cytosine and adenine. To read such a genome, it must be broken into millions of small pieces, which are read using electrochemical arrays. Each sequence is then carefully aligned and reassembled to provide a finished read.

The first complete human genome was sequenced between April 1999 and June 2000. The cost of that first draft of our genetic script was about $300 million. It needed subsequent refinement at a cost of a further $150 million. So to decode that first genome cost at least $500 million, and possibly as much as $1 billion.

But this changed fast. By September 2001, the National Human Genome Research Institute of America's National Institutes of Health reckoned that the cost of sequencing a full human genome ran to $95,263,702. By August 2019, that price had declined to $942, a 100,000-fold improvement.[9] Illumina, an American firm, was behind this – it's HiSeq X sequencer allowed sequencing costs to break the $1,000 mark in 2014.

Then, after a few years of dominance by Illumina, the Shenzhen-based company BGI announced in March 2020 it was capable of sequencing a full genome for only $100 – representing a million-fold improvement in less than 20 years.[10] That equates to a halving in price every year for two decades. It puts Moore's Law to shame. If the price-performance improvements in genome sequencing had followed the same curve as Moore's predictions for microchips, at the time of writing you could expect to pay more than $100,000 per sequence. In fact, genome sequencing has fallen 1,000 times further than Moore's Law would have forecast.

And, as ever, lower prices mean higher usage. In 1999, we had sequenced one messy genome. By 2015, humanity was sequencing more than 200,000 genomes a year.[11] One research group estimates that by 2025 as many as 2 billion human genomes may have been sequenced.[12]

There are a number of factors driving down the cost of coding a genome, and growing computing power is part of the story. Genome sequences are enormous chains of letters. Coding one human genome would require about 100 gigabytes of storage (enough to store about 25 high-definition movies) – a level that is much easier to secure now than it was two decades ago. But Moore's Law is far from the only cause of the price drop. There have been developments in the way we produce the reagents and 'amplifiers' required to turn a DNA sample into something readable. Over the years, these chemicals have become progressively cheaper and cheaper. Meanwhile, advancements in electronics have allowed scientists to create cheaper sensors, and developments in robotics have allowed for more automation of the manual parts of the convoluted process.[13]

Genetic coding is but one aspect of the revolution in biotech. Another is synthetic biology, a field that melds several disciplines, including computer science, biology, electrical engineering and biophysics, to create novel biological components and systems. It too is on an exponential march – one that is churning out breakthroughs in agriculture, pharma, materials and healthcare. Today, we can sequence and manipulate micro-organisms. We can turn them into little natural factories to produce the chemicals and materials we need – something that would have been unthinkable even a decade ago. The impact of this will be transformative. According to some estimates, 60 per cent of the physical 'inputs' in the global economy could be produced biologically by 2040.[14] Harnessing nature in this way will let us produce completely new materials – biopolymers that won't hurt our oceans; electronic components that are lighter or consume less power.

And then there is the last of our four areas of exponential change: manufacturing. How we make things is in the process of transforming fundamentally, perhaps for the first time in millions of years.

We have been manipulating physical materials – matter – in pretty much the same way since long before there were *Homo sapiens*. The oldest-known flints were carved in Olduvai, in modern-day Tanzania, about 1.7 million years ago. Industrial-era manufacturing processes have much in common with those of our distant forebears. We too generally use a subtractive process – start with a block of stuff, and chisel away what we don't want. This is what hominids did with flints. It is what pharaonic stonemasons did to the stone blocks of the pyramids. And it is what Michelangelo did when he chiselled a block of marble to create *David*.

Today, we can do this at a grander scale and with greater precision, but the process is essentially the same. Even as the computer age heralded precise computerised machining, this was still a subtractive process: the early human's hammering of flint on stone was replaced by a diamond cutter controlled by a computer. Of course, there are other methods of making stuff, such as using casts to mould metals or plastic. These have the advantage over chiselling in that they don't waste any material. But a big disadvantage: casts and moulds only create copies of a single design. Want a new product and you need a new mould.

Additive manufacturing, or 3D printing – I'll use the terms interchangeably – is an exponential technology that delivers the individual detailing of subtractive manufacturing, without the waste. Typically, objects are crafted through computer-aided design. The process involves creating a new object from scratch: by putting together layers upon layers of melted material, using a laser or a device a little like an inkjet printer. The material can range from glass to plastic to chocolate. It marks a fundamental break with many millennia of subtractive manufacturing, and thousands of years of casting and moulding.

Since the first 3D printers were developed by Charles Hull in the mid-1980s, additive manufacturing has improved dramatically. The process has become faster, more precise and more versatile – today, 3D printers can work with materials including plastics, steels, ceramics and even human proteins. In 1999, the Wake Forest Institute for Regenerative Medicine grew the first 3D printed organ for transplant surgery. And in Dubai in 2019, my friend Noah Raford spearheaded the then largest 3D-printed object: a single-storey 2,500-square-foot building.[15] Printed out of concrete in 17 days, Noah used it as his office for several months. The project used 75 per cent less concrete than a typical design and was built with unheard-of precision.

Additive manufacturing is still a tiny business. You'll find it in prestige products and in highly specialist sectors of the economy – lightweight parts for fighter jets, or medical implants. But the underlying technologies are on an exponential course. Researchers estimate that most additive manufacturing methods are developing at a pace of between 16.7 per cent and 37.6 per cent every year, with the average rate falling in the high thirties.[16] Over the next 10 years we will see performance improve 14 times – and, of course, see prices drop concomitantly. Terry Wohlers, an analyst of the additive manufacturing sector, tells me that the 3D printing market grew 11 times in the decade to 2019 – a rate of 27 per cent per annum.[17]

Why does the transformation of these four domains matter? After all, new technologies come along all the time. Researchers develop new ways of solving problems; engineers improve the methods we already have; occasionally we stumble upon a breakthrough. Even if the rate of change is increasing, you might think, the fundamental process is nothing new.

But the technologies in these four domains – computing, biology, energy and manufacturing – are special. To understand why, we must recognise a fundamental truth about innovation: not all technologies are equal.

Most technologies have fairly few uses – think of stirrups, or light bulbs. This doesn't mean their impact is small. The humble stirrup, really only of use for the horse rider, is credited with helping Genghis Khan sweep across Asia to create the world's largest land empire. The light bulb broke us free from the shackles of darkness. Society could function, at work or at home, after sunset. Narrow technologies can therefore have a broad impact. However, their uses remain relatively circumscribed.

Some innovations have a much broader utility, though. The wheel might provide power in a waterwheel, or serve as part of a pulley, or be used in a vehicle. Farmers, firemen and financiers might all have cause to call on a wheel. And a wheel could be used in every part of their trade. These wide-ranging inventions are known as 'general purpose technologies'. They may displace other technologies and create the opportunity for a wide variety of complementary products – products and services that can only exist because of this one invention.

Throughout history, general purpose technologies (GPTs) have transformed society beyond recognition. Electricity drastically altered the way factories work, and revolutionised our domestic lives. The printing press, which played a key role in the European Reformation and the scientific revolution, was much more than a set of pressure plates and cast metal type. GPTs upturn our economies, and our societies too – spawning changes far beyond the sectors in which they began.[18] As the economists Richard Lipsey, Kenneth Carlaw and Clifford Bekar put it, GPTs 'change almost everything in a society . . . by creating an agenda for the creation of new products, new processes, and new organizational forms.'[19]

Part of the reason GPTs are so transformative is the way they have effects beyond any one sector. Consider one of the key GPTs from the start of the twentieth century: the car. To reach their potential, cars needed suitable roads – physical infrastructure that spanned nations. But cars also needed fuel and spare parts, and drivers needed sustenance – creating the demand for fuel stations and roadside cafés. Cars forced changes to the urban environment and so cities started to change, with precedence going to motorised vehicles. Over time, suburbs developed, and with them came the gradual reshaping of consumer practices: reasonably priced hotels for vacationers, and big-box retail stores. New rules slowly emerged, including a slew of safety regulations for drivers. In short, the GPT changed everything.

And this hints at why the exponential revolution in our four key sectors is so important. We are witnessing the emergence of a transformative new wave of GPTs. Not a single GPT, as in the time of the printing press. Nor even three GPTs, like the early twentieth century's offering of the telephone, the car and electricity. In the Exponential Age, we're experiencing multiple breakthrough technologies in the four broad domains of computing, energy, biology and manufacturing.

At this early stage, it is hard to pin down precisely what the GPTs in these areas will be. All we know is that the emerging technologies in each sphere can be applied in a massive array of ways. As we have seen, growing computing power has a seemingly endless range of uses. Gene engineering might be used to tamper with microorganisms, produce new screens for smartphones, or help us design precise medicines. And 3D printers let us create everything from precision car parts to new bodily organs.

The rapid evolution of these technologies does not presage instantaneous change. The revolutionary effect of general purpose technologies may take time to materialise. Consider electricity. In the words of the leading economic historian James Bessen: 'The first electrical generating stations opened in 1881, but electrification had little effect on economic productivity until the 1920s.'[20] GPTs take a while to have meaningful effects – as new infrastructure is built, ways of working change, and companies train their employees in novel techniques.

GPTs are integrated into the economy in a series of steps, best described by the economist Carlota Perez.[21] First comes the installation phase, when the basic infrastructure underpinning a GPT is developed. The roll-out of a GPT is an arduous process: to build an electricity network, you need to build generating systems and power lines and grids. In this phase, skills are limited, and know-how is scarce. It takes time to develop the knowledge required to apply a GPT at scale. These early years may be a low-productivity, discovery-led period in the life of a GPT. Existing, mature technologies are more efficient and pervasive than the new inventions – in some cases, using a novel technology may be more trouble than it is worth.

But the real revolution comes at the next stage: the deployment phase. After a laborious installation phase, our economies have rolled out enough of a new technology to do something useful with it. Companies have figured out its purpose. Crucially, managers and workers have built up the requisite knowledge and experience to make use of the technology. And complementary services, like repair and supply shops, have been established. During the deployment phase, societies can enjoy the boons of a GPT: this is a 'golden age', in Perez's words. But it takes time.

As we saw in the last chapter, however, today's technologies – general purpose or otherwise – are rolling out at a much faster rate than ever before. Much of the infrastructure, from cloud computing to smartphones, has already been deployed. And that means that the transformation Perez alludes to may come faster than in any previous era. Our age is defined by the cascading of technologies: one new technology leads rapidly on to the next, and on to the next.

Consider the wide-ranging effects of computing – which stretches well beyond the traditional parameters of IT. As Perez points out, the computing revolution made the creation and processing of information much easier. One consequence was the rise of products that could target increasingly small niches, from many flavours of ketchup to a dazzling variety of clothes – enabled by our enhanced ability to gather and process data. Another was the transformation of corporate organisation: firms changed the way they operated and developed different strategic objectives, for reasons we'll encounter in Chapter 4.

But the rise of computers was only the first step. The spread of the PC laid the foundations for the rapid spread of the internet. In 1984, there were 1,024 computers connected to the internet. By 1994, the number had risen more than 3,000-fold. One estimate put the number of internet users in 1995 at 16 million; by the end of 2020, more than 5 billion of us were connected to the web (a 300-fold increase).[22]

And the spread of the internet was facilitated – and in turn accelerated – by the rise of the smartphone. While the first smartphones emerged in the 1990s, they were simply not up to the task. Battery life was poor, screens were small, and they ran on very limited software. IBM's Simon, released in 1992, only sold 50,000 units.[23] But everything changed with the launch of Apple's iPhone in 2007, the first breakthrough smartphone. Within five years, a billion of us had one of these all-purpose devices. By the end of 2020, that number had reached 3.5 billion.

As these innovations interact, they transform numerous parts of the economy. The smartphone has replaced many other consumer devices – Walkmans, calculators, diaries, watches and street maps. Camera sales have collapsed in the wake of the rise of phone cameras. And shopping has changed too. American consumers spent $284 billion shopping via their mobile phones in 2020.[24] In November 2019, during an annual promotion called Singles' Day, China's shoppers placed orders totalling $39 billon to Alibaba through their smartphones. A year later, the same promotion netted $74 billion in sales. Retailers who had invested in expensive storefronts in costly areas found that customers were happier lying in bed prodding a screen than visiting physical shops.

This is the power of GPTs. Their effects spread inexorably from area to area, rippling across all aspects of our daily lives. And the GPTs of the Exponential Age are only just beginning to emerge. We have so far witnessed the rise of personal computers, the internet, smartphones; we haven't yet experienced the effects of extremely cheap power, bioengineering, 3D printing – and many other technologies too nascent to predict.

So far, we've explored what some of the new generation of exponential technology consists of. And we've learnt why these innovations – many with the hallmarks of general purpose technologies – will have such a transformative impact. But this all raises a new, even bigger question: why now? In other words, what is driving the exponential revolution?

The answer lies in three forces, which the rest of this chapter will focus on. And the first can be summarised with a simple maxim: 'Learning by doing.'

To make sense of this first cause, we need to return to Moore's Law – and come to understand its limitations. There was always something that bugged me about the theory: its relationship with time. As outlined by Gordon Moore, the passing of time is key: every two years or so, you can get twice as many transistors on the same-sized silicon wafer for the same price. When I first came across Moore's Law in the mid-1980s I took it at face value, as any teenager might. It was alluring and easy to remember. But my entry into the world of tech as a professional led me to ponder it more critically. What was it about the passage of time that magically packed more semiconductors onto silicon? Surely humans had some agency in this process? Consider a strike. What if workers downed tools, locked up the fabs and picketed outside their factories for two years? Would the transistors still get smaller? Surely not.

As we saw in the last chapter, Moore's Law is a social fact, willed into existence by industry. And that means human behaviour is key: if we stopped endeavouring to make Moore's Law true, it would cease to be true. As such, while it is an adequate description of technological change – for now, at least – it is ill-equipped to explain why technologies improve.

A better approach would be a law of technological change that is based not on time, but on what actually happens in a sector. Such a model could account for changes in our behaviour: if we stopped building microchips, it wouldn't immediately lose its explanatory power. And ideally, we would want such a prediction to be applicable to whole families of different technologies, not just to one industry.

Fortunately, we have just such a principle – and it was theorised two decades before the silicon chip was invented. Wright's Law was developed by Theodore Wright, an aeronautical engineer who set out to understand how much aircraft cost to manufacture and why. He looked at the production costs of planes in the 1920s and 1930s and noticed that the build cost seemed to decline following a pattern. The more aircraft were built, the more airframes the engineers, mechanics and designers had to assemble – and the cheaper each unit became.[25]

His theory was that for every doubling in units produced, costs would fall by a constant percentage. The exact nature of the decline would depend on the engineering in question. In the case of the aircraft Wright studied, it was a 15 per cent improvement for every doubling of production. This 15 per cent improvement is known as the 'learning rate'.

In Wright's analysis, the reason was simple. As engineers build a product, they come to understand what it takes to build it better. They figure out a more elegant way to connect two components. Or they combine a set of different elements into a single component. Workers figure out shortcuts that make them more efficient. In other words, they learn by doing. As engineers finesse the process, a small innovation here and another innovation there drives rapid increases in efficiency.

For this reason, the key to the continuation of Wright's Law is increasing volume. Greater demand drives improvements in the process, which in turn drive down costs, which in turn drives further demand, and so on. This is a distinct concept from the notion of economies of scale – the idea that efficiencies come from having bigger operations, or getting better prices from suppliers. Rather, Wright's emphasis is on the relationship between demand and skill. As demand for a product grows, the people producing it have to make more of it. And that means more opportunities to learn by doing. As they put what they have learnt into practice, costs get driven down further and further.

This means Wright's Law has an edge over Moore's Law. Both describe the way the cost of technology diminishes exponentially. But Moore's Law simply describes performance improvements over time. There are scenarios it can't account for – those striking microchip-factory workers, for example. Wright's Law, meanwhile, connects progress to the quantities produced. Say that for every doubling of production, the unit cost of a gadget drops 20 per cent. If production doubles every two years, costs will drop by 20 per cent every two years. If production doubles every year, costs will drop by 20 per cent every year. Wright's Law holds true even during that mythical strike: if production stops, the reduction in cost stops.

Since Wright's time, researchers have discovered that his law applies to dozens of technologies – from the outputs of the chemical industry, to wind turbines and transistors. The doubling of volume does lead to relatively constant declines in per-unit price. And this holds true for the hallmark technologies of the Exponential Age. The declining cost of lithium-ion batteries has been successfully predicted using Wright's Law. Electric car sales increased 140-fold between 2010 and 2020, and every one of those units needed a lithium-ion battery. In the same ten-year period, demand for such batteries shot up by a factor of 665 (the batteries were getting more capacious each year). The increase in volumes led to a decline in the per-unit price: from 2010 to 2020, the price of an average battery pack fell by nearly 90 per cent. But it's not just batteries. As it turns out, Wright's Law even more accurately describes what happened to silicon chip prices than Moore's Law does.[26]

This law has been with us since long before the Exponential Age. But today, there's one crucial difference. Historically, Wright's Law has had clear limits. Think of the way a product's entry to the market follows an S-curve. At the beginning, it spreads exponentially – but as the market saturates, uptake tails off. This was assumed to hold true for Wright's Law too: eventually, the market saturates, and the decrease in prices slows to a halt. Wright, who died in 1970, might have been disgruntled to discover what ultimately happened to the prices of the aircraft he studied. The original Boeing 737, first built in 1967, cost $27 million in 2020 terms. The latest variant, the 737 MAX which first flew in 2016, cost as much as $135 million – five times more. So much for price declines.

Yet a striking feature of our age is that the hard limits to Wright's Law seem much more distant – and in some cases they might not even exist. Today, the prices of new technologies seem able to drop endlessly. We've already seen this process in action a few times, so here let's home in on just one, particularly striking, example. Many of us are familiar with USB sticks, those little dongles used to move computer files around. These first appeared in the year 2000: $50 bought you 8 megabytes of storage.[27] Twenty years later, $50 could buy you a good-quality 2-terabyte flash drive (storing 250,000 times more of your memories). This represents an annual increase of 85 per cent. For an industrial-era product – say, a washing machine – at some point in this process Wright's Law would have tapered off. For USB sticks, there has been no such slowdown.

Why do the limits to Wright's Law seem so much more distant in today's economy? To some extent it relates to the physical nature of the underlying technologies, which are fundamentally different to older inventions. Silicon chips get faster as you make their components smaller. And since chips live on a square wafer, each time you shrink a component you get the efficiency gain squared. If you have a wafer that is 100 square millimetres and you can fit one component onto every millimetre, then a single wafer can hold 10,000 components (100 x 100). If you shrink the components by 50 per cent, so that you can fit two onto every millimetre, you end up with 40,000 components (200 x 200) on the same die.

This process holds true for many exponential technologies. Even mighty wind turbines are not immune to such effects. The generating capacity of a wind turbine is proportional to the area that the blades sweep through. That area grows by the square of the length of the blades, so if you are able to manufacture a blade twice as long, you get quadruple the impact.[28] In 1990, a typical large wind turbine had a trio of blades with a diameter of 40 metres and was capable of generating 0.5 megawatts of power.[29] By 2020, General Electric was delivering wind turbines with 24 times that output and a diameter of 220 metres – 5.5 times longer than the 1990 vintage.[30]

But the greater cause of the newfound power of Wright's Law lies in economics. Previously, the S-curve of demand tapered off when a market reached saturation. Today, that point of market saturation is much more distant – because global markets are so much larger. And this means that the process explained by Wright's Law can continue for much longer, and the exponential gains can continue to mount up.

As we'll see time and again in this book, the rise of a global market for products is one of the great changes of the last 50 years. The volume of world trade increased 60-fold, from $318 billion to $19,468 billion by 2020. And this pushed us ever further down the path of Wright's Law. Bigger markets mean more demand; more demand means more efficient production; more efficient production means cheaper goods; cheaper goods mean bigger markets. The cycle has an inherently exponential logic. And so Wright's Law describes how technological progress takes on its own momentum: the more we make of something, the more demand there is, and so the more we make.

This, at heart, is why Wright's Law is driving us into an exponential future. We learn by doing. And in recent years, we have been doing more.

The second driver of exponentiality is even simpler: combination. The general purpose technologies of the Exponential Age are not only improving at exponential rates, they are also combining in novel and powerful ways. Today's GPTs riff off each other in unpredictable, constantly shifting patterns. And as these novel uses of technology are pioneered, they help other technologies evolve in fresh directions.

One good example of the power of combinatorial invention comes from the work of Bill Gross. He is trying to help decarbonise the economy by building new systems of energy storage – his attempt to solve the storage problems facing renewable energy that we encountered earlier. His company, Energy Vault, is building a huge electricity storage system – enormous, insect-like cranes with six arms radiating out from a central tower, and gigantic blocks of reconstituted building rubble. The cranes, powered by locally sourced solar power, are used to hoist the cuboids into the air. As they are stacked, much of the electrical energy used to raise these massive slabs is converted to gravitational potential energy.

When the blocks are stacked, the tower forms a battery storing the equivalent of 35 megawatt-hours of electricity from its gravitational potential – enough to fully charge 1,000 compact electric cars or power a typical American home for more than nine years. When power is needed at night, because it is dark and the solar panels are dormant, the cranes can lower the blocks. This converts the gravitational potential energy into kinetic energy, which in turn drives electrical generators – spitting much-needed current into the electricity grid.

What does this have to do with combination? Well, it is only possible thanks to the interaction of several different technologies. This giant crane-block battery depends on a particular mix of four, very well-understood technologies – the cranes, the building aggregate, the generator which converts the dropping of blocks to energy, and the shipping systems that let us move these things around. And then there's a fifth, more unexpected technology: an automated 'machine vision system' using deep learning. Each crane has a set of cameras whose input is automatically processed by a computer. This computer controls the cranes and the lifting and placing of the blocks. It obviates any human operators – and it is the absence of a human operator that allows Energy Vault to hit a competitive price.

Of course, technologies have always combined. Flywheels and cranks were combined to create water pumps in medieval Europe. The light bulb involves electricity, glass bulbs and the harnessing of inert gases. And this has been even truer of general purpose technologies. Henry Ford's production system was made possible thanks to the emergence of one GPT, electricity, which allowed him to build another GPT, the automobile. Invention is a little like a family tree. New breakthroughs rely on aspects of the technologies that went before.[31]

But combination has never taken place at the scale it does today. For several reasons, contemporary technologies are more prone to combination than those of the early to mid-twentieth century. One factor is standardisation. Today, standard components can be used in different compound products. Think of an AA battery – you can use it in your remote control, electric toothbrush, torch or toy car. This standardisation makes it easier and cheaper to build more complex products. The toothbrush maker doesn't need to be an expert on batteries; they can simply buy in the technology. And so standardisation increases the speed with which a product can go to market and scale.

Standardisation is, on one level, a step up from an old idea: replaceable parts, a military innovation of the eighteenth century. This allowed soldiers to carry spare parts that could fix any musket; before then, each rifle had needed specialist components. But today, standardisation is easier and more widespread than any musketeer could have imagined. In recent decades, international standards bodies such as the International Telecommunication Union and the International Organization for Standardization have taken the logic of replaceable parts to new heights. These groups build consensus around simple things (like components) and complex things (like manufacturing processes). Once a standard exists, it eliminates headaches. Electrical sockets do not adhere to an international standard. It is why we travel with adapters and, in many cases, find ourselves stuck with the wrong plug for our destination.

But the most important standards of the Exponential Age have not been imposed from the top down by international organisations. Rather, they have been developed from the bottom up, often by the very people who are creating new technologies. The internet is built on standards created in this manner, developed by the academic researchers who stewarded the network until the early 1990s. The email protocol – the set of rules describing what an email is, and how sending and receiving computers should process it, is described in two documents: the RFC (Request for Comments) 821[32] and the RFC 822.[33] Both date from 1982, and were written by the late Jon Postel of the University of Southern California and David Crocker, then of the University of Delaware, respectively. The web protocol became established as a de facto standard within a couple of years of its development by Tim Berners-Lee in 1989. These internet standards mean we can send emails from one person to another without worrying about the compatibility of our email systems. They are, in short, interoperable. And we all benefit hugely from that interoperability.

Standardisation of this kind makes the world more efficient – and allows different innovations to combine. When technologies adhere to standard forms, they can be applied to a wider range of industries. Standard technologies become like Lego blocks that can be selected and assembled into an eclectic array of services. This combination and recombination catalyses further innovation.

I have experienced the transformative recent effects of the new age of standardisation first-hand. In 2006, I ran an innovation group at Reuters, the news and information company. The then CEO, Tom Glocer, was always keen to venture into new frontiers. He put me in charge of realising his latest scheme: to get hold of regular satellite photographs of major ports – Singapore, Shanghai, Busan, Rotterdam – and use them to develop economic insights for our clients. Busy ports would presage a booming economy; fewer tankers and half-laden freighters might herald a slump.

A brilliant idea. But a hard one to realise. It wasn't really clear where you start with a request like that. I took to calling the switchboard at NASA and the European Space Agency. I know now, as a result of the research for this book, that there were only a handful of commercial earth observation satellites in orbit, mostly with the wrong kind of sensors for what we wanted to achieve. I did not, alas, know this at the time. And I failed Tom.

But today, it would be trivial – and cheap, measured in tens of dollars – to give Tom what he wanted: accurate, up-to-date photographs of almost any part of the planet. Such images are used by hedge funds and commodities traders in precisely the way my boss envisaged. These companies count cars in shopping-mall parking lots to estimate customer demand, and ultimately assess how well the retail sector is going. Or they analyse the shadows cast by oil tankers to estimate their load – and hence the global demand for oil. Lemonade, an American insurance company, uses satellite imaging to estimate the risk of forest fires before it offers home insurance policies.[34]

By the end of 2018, there were about 2,000 operational satellites spinning around the globe. According to the Union of Concerned Scientists, between 1991 and 2000 there were 118 satellite launches, about 10 a year – four-fifths of which were commercial, mostly for communications (the remainder were governmental or military).[35] In 2018, 372 satellites were launched, more than 20 times the level at the turn of the century.

And the rise of this new generation of satellites is, you guessed it, thanks to standardisation. Standard components have made satellites more affordable and space more accessible. Standard components enable standard-sized satellites – like the dinky low-orbit satellites known as 'CubeSats'. Rockets then accept standard-sized satellites, dramatically reducing the costs of getting out of the earth's atmosphere.[36]

In software, standardisation has become commonplace. It has given rise to a componentisation: many frequent tasks we ask of software are now available as easy-to-access lumps of code. A modern software developer may spend as much time sticking together such standard components as they do writing something new. A new mobile app might involve tens of thousands of lines of code or more, but the developers writing the apps don't type that code themselves. Rather, they will find standardised components developed by others. Such commonality means firms can draw upon expertise from many different industries: someone developing a travel-booking app might use a calendar component originally developed for a meeting-scheduling app.

All this points to a shift in the nature of technological development. Modern technologies are more likely to combine and recombine. And this process of combination leads to the emergence of new innovations.

Laura O'Sullivan was finishing high school at the Mount Mercy College in Cork, Ireland, when she decided to develop an automated system that could detect anomalies in cervical smears. The previous year there had been a cervical testing scandal in which more than 200 Irish women had wrongly been given the all-clear only to develop cancer later. Laura, then 16, resolved to tackle the problem. She was aware of recent breakthroughs in machine vision – the ability of computers to identify objects or patterns in images that we encountered in Chapter 1. Perhaps, she thought, this technology could be used to better identify malignancies in cervical smear images.

Laura had no more than rudimentary coding experience – a couple of holiday programming camps, but no formal training. 'I took some online courses on Coursera and from Stanford, on machine learning and deep learning. I needed to understand the basics of how this worked,' she told me. She started her project over the summer holidays, learning how to build and fine-tune convolutional neural networks and how to find and clean the data. Fortunately, Herlev Hospital in Denmark had an open-source data set of cervical smears she could use.

It wasn't straightforward. The data set was, in the jargon of a data scientist, unbalanced. It contained too many supposedly abnormal, potentially cancerous screens, and not enough healthy ones. Real-world data would be the reverse: most women have healthy smears, and a tiny number have problematic ones. This kind of inconsistency, or lack of balance, in the data Laura was using could cause problems for her system.

So she found a way to artificially create more data, representative of healthy specimens. The effect of this, done carefully, was to create a robust data set that would allow the algorithms to learn effectively. The technique she used, generative adversarial networks (or GANs), was red-hot. The first results from GANs had been established by Ian Goodfellow, a researcher based in California, only four years earlier.[37] Laura was able to download the code which would run the GANs for free from GitHub, a website where software developers collaborate and share their software freely. All the computing was done on her dad's consumer-grade PC. This had one adornment Laura loved: two screens, meaning she could look at her code on one and the computing tutorials on the other.

By December 2018, Laura was fine-tuning her results. When I met her in January 2019, at the finals of the Irish Young Scientist Competition, her new software was doing better at identifying anomalies in images than a human doctor. Unsurprisingly, she won a prize.

Laura's experience is a perfect example of the third driver of exponential technology: the profusion of networks. Over the last 50 years, there has been a proliferation of networks of information and trade. It has never been easier to send funds from one part of the world to another. It has never been easier for a meme to travel from Santiago to

Sydney.[38] It has never been easier to transport a piece of electronic equipment from Shenzhen to Stockholm. And, for that matter, it has never been easier for a virus to spread from a remote location in a vast country to 100 nations in a matter of weeks.

Networks have changed the nature of trade, invention, science, relationships, disease, finance, information, threats and more. And, crucially, these flows lead to the exponential development and spread of technology.

We can identify a few different forms of network that are particularly important in catalysing exponential technology. First, networks of information. Informational networks have been developing for decades. During the 1970s, as computers became more commonplace in academia, it made sense to use them for sharing research. About the same time as the growth of academic computing came the expansion of the internet. By 1990, 300,000 computers in more than a dozen countries – mostly in universities – were connected to it. Scientists could easily email their papers to each other. And they did.

From there, it was a simple step to create big, free, academic databases. Paul Ginsparg, a young physics researcher, was struggling with the flood of emailed papers. He dreamt up a centralised system where all preprints could be downloaded. When I first accessed it in mid-1992, it was hosted at the Los Alamos National Laboratory. Researchers reached it via a programme called Gopher – a precursor to the World Wide Web – at the alluring internet address xxx.lanl.gov. Today, Ginsparg's creation is known as arXiv (pronounced 'archive': the 'X' represents the Greek letter *chi*). And it has revolutionised the distribution of scientific knowledge.

Good-quality papers, though not peer-reviewed ones, were available by the thousands on arXiv. Ginsparg and his collaborators ignited the 'open access' movement that seeks to widen the availability of scholarly thinking. Today, arXiv has expanded from its start in high-energy physics into disciplines like astrophysics, computer science and maths. More than 10,000 papers were submitted to arXiv by 1994; in December 2019, 23 million papers were downloaded across all disciplines.[39] As every one of these papers has always been available on arXiv before making it into a print journal with its stringent peer review, the moniker 'preprint server' has stuck.

This phenomenon is not limited to physics: other domains have seen the benefits of the preprint. BioRxiv was launched for biologists in 2003, PsyArXiv for psychologists in 2016, and SocArXiv for the social sciences in the same year. Around 50 preprint services now work to accelerate the diffusion of academic knowledge.[40]

These preprint servers are so powerful because they remove the boundaries to academic research. They let ordinary people access cutting-edge ideas for free. And that widens the groups who can participate in the scientific process. This was never truer than in the response to the coronavirus pandemic. The first academic paper on the virus was published on a preprint server on 24 January 2020. By November 2020, more than 84,000 papers about Covid across disciplines were available on preprint servers and other open-access sources.[41] And Laura O'Sullivan is another example of the power of the preprint server in speeding up the spread of new ideas. It is astonishing that an idea as powerful as generative adversarial networks can travel from a non-peer-reviewed white paper to a high-school project a continent away, all in less than five years.

Preprint servers are but one example of how new information networks – often online – become tools for rapid collaboration. The internet enables collaboration through thousands, millions, of other networks. GitHub lets 56 million software developers collaborate across 60 million different software projects;[42] Behance helps designers collaborate on creative projects; blogs allow people to write about and comment on ideas; Wikipedia allows anyone to access and contribute to an ocean of accessible, specialist information. Social networks – often online – perform a similar function, as do private chat groups.

These information networks catalyse the emergence of exponential technology. They spread the know-how that makes breakthrough technologies possible. The internet distributes ideas across the globe to millions of willing recipients.

While informational networks help ideas spread worldwide, the shipping container does the same for physical products. In the 1950s, commodities like oil and grain tended to come in tankers or bulk freighters; other goods were shipped in a variety of different-sized boxes or carried loose on ships. Stevedores would manually load and offload cargo. It was a cumbersome, slow and expensive process. Moving goods from the US to Europe or vice versa might take three months, and shipping costs might run to 20 per cent of the cargo's value.

Containerisation changed all that. The first shipping containers arrived at the Port of Houston on the *Ideal-X* in April 1956. The containers were 35 feet long, 8 feet wide and 8 feet high. Here, too, standardisation helped: by 1965, the International Standards Organisation had agreed a specification for shipping containers (8 feet wide, 8½ feet high, and 10, 20 or 40 feet long). Containers could be handled by ports the world over and rolled straight onto flat-bed trucks. And in March 1966, the first ships holding such containers departed the US.[43]

That first ship held 226 containers. A year later, a container ship holding 609 containers started to ply the seaway between Oakland on the west coast of the US and Cam Ranh Bay, an American military base in Vietnam. And these ships only got bigger. By 2020, the world's largest container ship, the HMM *Algeciras*, fully laden, could hold 23,964 of the boxes.[44]

The effects were staggering. Between 1980 and 2015, the global capacity of container ships worldwide grew 25-fold. This was part of a long-term trend, reducing the cost of doing business globally. Between the end of World War Two and 1980, when container ships comprised less than a tenth of global cargo fleet tonnage, sea-freight costs declined 2.5 times. Over the subsequent thirty years, container ships would account for three-fifths of a much larger global cargo fleet – and prices would halve again. Volumes dramatically increased with declining prices. Between 2000 and 2018, container port traffic more than tripled.[45]

These new shipping technologies combined with the internet, e-commerce and computerised ordering systems to make logistics increasingly efficient. The 'just in time' supply chain means that much of what we might buy in a week currently sits on a shipping freighter on an ocean somewhere. Huge behemoths like Apple hold fewer than ten days of stock, or 'inventory outstanding'. Millions of dollars of products that the company knows it will sell in two weeks' time have not yet been produced. It can rely on the digital network of the internet to take the order and coordinate it, and a physical network of trucks and ships to meet all foreseeable demand.

Such trade networks have contributed to the inexorable development and spread of new technologies. They allow products to diffuse into far-flung markets faster than ever before. New products – the latest phones, the latest laptops – can launch worldwide on the same day. When Apple launched its first iPhone on 29 June 2007, the device was available for sale in a single store in San Francisco. By the time the eleventh version of the phone was launched on 20 September 2019, it was delivered to eager shoppers in hundreds of different cities in more than 30 countries around the globe.

These networks, then, are the third engine of the Exponential Age. As the world gets more connected – by information and by shipping lanes – it becomes easier for technology to span the world at pace.

By now, we can start to make out where the new wave of exponential technology has come from. Exponential technologies are being driven by three mutually reinforcing factors – the power of learning by doing, the increasing interaction and combination of new technologies, and the emergence of new networks of information and trade.

But for this picture to be complete, we also need to understand the economic and political context. As you may have noticed, none of these driving forces can be explained without reference to a wider set of shifts in politics and economics. In particular, the process of globalisation. Our continual ability to learn by doing is based on the ever-growing international demand for goods. The standardisation of parts, software and technology protocols is partly down to the emergence of global standards bodies. The rise of trade networks has as much to do with the development of new markets as with the advent of new technology. Technology, politics and economics are intertwined. In fact, we can trace the mutual reinforcement of exponential technology and globalisation closely. Moore's Law was expounded in 1965. And the first international container ship departed from New Jersey in March 1966.

So, we need to look beyond technology to understand the origins of exponential change. Just as the hallmark technologies of the Exponential Age were nascent, yet to really make their mark, a new political orthodoxy was coming of age. According to the historian Binyamin Appelbaum, it was in the late 1960s and early 1970s that economists in the US began to 'play a primary role in shaping public policy'.[46] Economists had hitherto lived on the peripheries of the academy and in the bowels of the central banks. But in the 1970s, everything changed.

It's a well-known story. Rich economies were assailed by a toxic combination of low growth and high inflation – known as 'stagflation'. Waves of strikes and fuel crises eroded trust in Western governments. And, as faith in government reached a new low, voters and policymakers cast around for a new way of doing things. The scene was set for the emergence of a new school of economics, whose standard-bearer was University of Chicago professor Milton Friedman.

Friedman's belief was that markets would work better if the government got out of the way. Since the end of World War Two, Western governments had adopted a fairly interventionist approach to their economies. Until 1979, the top rate of income tax, payable on investment earnings, was 98 per cent in the UK and 70 per cent in the US. In the UK, the basic rate of tax was 33 per cent by the latter half of the 1970s. And with high taxes came hands-on government: nationalised industries, heavy regulation and interventionist industrial policy. But Friedman's acolytes advocated a different approach. By rolling back regulations and cutting taxes, governments could unleash the power of the market – and bring the return of high growth and manageable inflation. This market-friendly ideology found its touchstone with Friedman's famous doctrine, which held that the social responsibility of corporations and the business sector was to increase profits, and not much else.[47]

The consequences were seismic. By 1976, when Friedman won the Nobel Prize in Economics, the University of Chicago was arguably the most important economic institution in the world. A wholesale embrace of the power of markets was in the making. With the election of Ronald Reagan in the US and Margaret Thatcher in Britain, these ideas found their way into government. Now the emphasis was on the market, not the state. In 1981, Reagan himself famously noted that he believed 'government is the problem', not the solution. Both administrations sparked a bonfire of deregulation, stripping away perceived hurdles to businesses' success. Along the way, they unleashed a wave of entrepreneurialism on both sides of the Atlantic, with new markets emerging everywhere from banking to hospitality to tech.

Most readers will already be familiar with this transformation. Many will have lived through it. But rarely do we dwell upon its relationship to technology – and particularly the exponential technologies discussed in this chapter. Free-market economics helped drive globalisation. And globalisation helped kindle the Exponential Age. All this means we can trace the beginnings of an era of exponential technology not just to the discovery of Moore's Law or the creation of the seminal Intel 4004 processor in 1971. Just as important was the emergence of a new political orthodoxy. From the late 1970s onwards, free-market capitalism would unleash the power of exponentiality.

From this moment, the foundations of the transition to the Exponential Age were laid. But it would take a little while to get going. Giving an actual date for the start of the Exponential Age is tricky. It is not like the age of flight (kick-started by Orville and Wilbur Wright's flight at Kitty Hawk, 1903), or the atomic age (the Chicago Pile-1 prototype reactor, 1942), or the space age (Sputnik's orbit, 1957). Exponential change is a continual, smooth curve; there are no sharp, disjointed moments. And as we saw in Chapter 1, the change starts off barely perceptibly, speeds up deceptively gradually, and only eventually does everything take off.

At point A on the graph below, the world feels static. At point B, everything is disconcertingly fast.

I believe point A on the graph lies somewhere between 1969 and 1971, with the development of the internet and the microprocessor. Even 30 years later, exponential technology was still trundling along slowly. When I was building internet services in the mid-1990s, offline products remained much more important than online ones. To be sure, some Silicon Valley gurus had begun to see the potential. But the rest of the world did not see it: the technology change was nascent, but not yet turning the 'real world' upside down. The website I built for *The Economist* in 1996 had a fraction of the readership of the print magazine.

And yet, at some point after the 1990s, we did reach point B. By 2020, *The Economist* reached 900,000 readers in print, but had more than 25 million followers on Twitter.[48] To my mind, the tipping point at which exponential technologies began to really transform everything came in the second decade of the twenty-first century. In 2010, 300 million smartphones were sold; by 2015, annual sales reached 1.5 billion. Between 2014 and 2015, the global average cost for solar photovoltaic energy dropped below that of coal. In business, 2011 was the first time for decades that the world's largest company was not an oil company: Apple pipped ExxonMobil for a few weeks. Short of a drastic change in fortunes, Exxon was the world's most valuable firm for the last time in

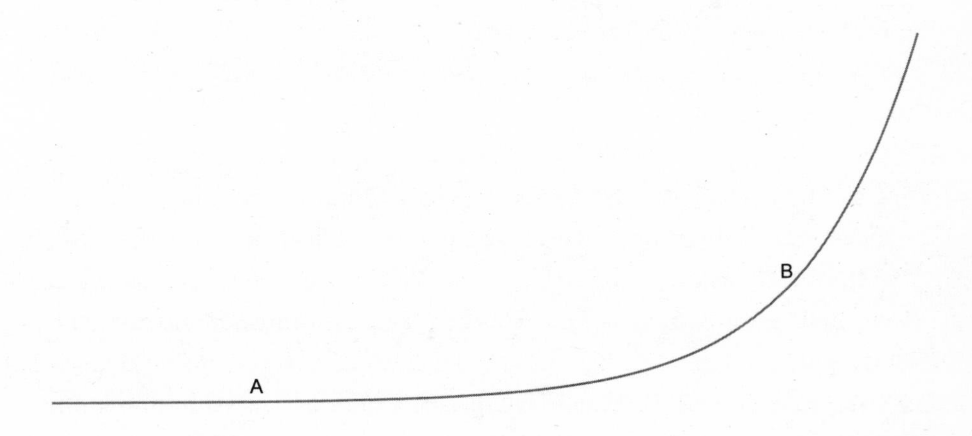

2013. By the beginning of 2016, six firms based on exponentially developing digital technologies – Apple, Tencent, Alphabet, Microsoft, Amazon and Facebook – were among the top-ten largest on earth.

This is what I mean by the Exponential Age. We are no longer just living in a period of exponential technologies – but at a time when these technologies and their effects are a defining force in our society. Like previous eras, the changes to our society are driven by the capabilities of new tech. But in recent years, these technologies have been developing at an unusually rapid rate – more than 10 per cent every year. As they improve, they combine and recombine, creating more and more possibilities. And they spill over into every area of our lives, rewiring our approach to business, work, politics – even our sense of self.

But the disruptive power of the Exponential Age does not just lie in exponential change itself. It lies in the way humans respond to this change. And it is this human response to the Exponential Age that we will examine next.

3

The Exponential Gap

In 2020, Amazon turned 26 years old. Over the previous quarter of a century, the company had transformed shopping. With retail revenues in excess of $213 billion, it was larger than Germany's Schwarz Gruppe, America's Costco and every British retailer. Only America's Walmart, with more than half a trillion dollars of sales, was bigger. But Amazon was by this time far and away the world's largest internet retailer. Its online business was about eight times larger than Walmart's. Amazon was more than just an online shop, however. Its huge operations in areas such as cloud computing, logistics, media and hardware added a further $172 billion in sales.

At the heart of Amazon's success is an annual research and development budget that reached a staggering $36 billion in 2019, and which is used to develop everything from robots to smart home assistants. This sum leaves other companies – and many governments – behind. It is not far off the UK government's annual budget for research and development.[1] The entire US government's federal R&D budget for 2018 was only $134 billion.[2]

Amazon spent more on R&D in 2018 than the US National Institutes of Health. Roche, the global pharmaceutical company renowned for its investment in research, spent a mere $12 billion in R&D in 2018.[3] Meanwhile Tesco, the largest retailer in Britain – with annual sales in excess of £50 billion (approximately $70 billion) – had a research lab whose budget was in the 'six figures' in 2016.[4]

Perhaps more remarkable is the rate at which Amazon grew this budget. Ten years earlier, Amazon's research budget was $1.2 billion. Over the course of the next decade, the firm increased its annual R&D budget by about 44 per cent every year. As the 2010s went on, Amazon doubled down on its investments in research. In the words of Werner Vogels, the firm's chief technology officer, if they stopped innovating they 'would be out of business in 10–15 years'.[5]

In the process, Amazon created a chasm between the old world and the new. The approach of traditional business was to rely on models that succeeded yesterday. They were based on a strategy that tomorrow might be a little different, but not markedly so.

This kind of linear thinking, rooted in the assumption that change takes decades and not months, may have worked in the past – but not anymore. Amazon understood the nature of the Exponential Age. The pace of change was accelerating; the companies that could harness the technologies of the new era would take off. And those that couldn't keep up would be undone at remarkable speed.

Such a divergence between the old and the new is one example of what I call the 'exponential gap'. On the one hand, there are technologies that develop at an exponential pace – and the companies, institutions and communities that adapt to or harness them. On the other, there are the ideas and norms of the old world. The companies, institutions and communities that can only adapt at an incremental pace. These get left behind – and fast.

The emergence of this gap is a consequence of exponential technology. Until the early 2010s, most companies assumed the cost of their inputs would remain pretty similar from year to year, perhaps with a nudge for inflation. The raw materials might fluctuate based on commodity markets; but their planning processes, institutionalised in management orthodoxy, could manage such volatility. But in the Exponential Age, one primary input for a company is its ability to process information. One of the main costs to process that data is computation.[6] And the cost of computation didn't rise each year; it declined rapidly. The underlying dynamics of how companies operate had shifted.

In Chapter 1, we explored how Moore's Law amounts to a halving of the underlying cost of computation every couple of years. It means that every 10 years, the cost of the processing that can be done by a computer will decline by a factor of 100. But the implications of this process stretch far beyond our personal laptop use – and far beyond the interests of any one laptop manufacturer.

In general, if an organisation needs to do something that uses computation, and that task is too expensive today, it probably won't be too expensive in a couple of years. For companies, this realisation has deep significance. Firms that figured out that the effective price of computation was declining, even if the notional price of what they were buying was staying the same (or even rising), could plan, practise and experiment with the near future in mind. Even if those futuristic activities were expensive now, they would become affordable soon enough. Organisations that understood this deflation, and planned for it, became well-positioned to take advantage of the Exponential Age.

If Amazon's early recognition of this trend helped transform it into one of the most valuable companies in history, they were not alone. Many of the new digital giants – from Uber to Alibaba, Spotify to TikTok – took a similar path. And following in their footsteps were firms who understand how these processes apply in other sectors. The bosses at Tesla understood that the prices of electric vehicles might decline on an exponential curve, and launched the electric vehicle revolution. The founders of Impossible Foods understood how the expensive process of precision fermentation (which involves genetically modified microorganisms) would get cheaper and cheaper. Executives at space companies like Spire and Planet Labs understood this process would drive down the cost of putting satellites in orbit. Companies that didn't adapt to exponential technology shifts, like much of the newspaper publishing industry, didn't stand a chance.

We can visualise the gap by returning to our now-familiar exponential curve. As we've seen, individual technologies develop according to an S-curve, which begins by roughly following an exponential trajectory. And as we've seen, it starts off looking a bit humdrum. In those early days, exponential change is distinctly boring, and most people and organisations ignore it. At this point in the curve, the industry producing an exponential technology looks exciting to those in it, but like a backwater to everyone else. But at some point, the line of exponential change crosses that of linear change. Soon it reaches an inflection point. That shift in gear, which is both sudden and subtle, is hard to fathom.

Because, for all the visibility of exponential change, most of the institutions that make up our society follow a linear trajectory. Codified laws and unspoken social norms; legacy companies and NGOs; political systems and intergovernmental bodies – all have only ever known how to adapt incrementally. Stability is an important force within institutions. In fact, it's built into them.

The gap between our institutions' capacity to change and our new technologies' accelerating speed is the defining consequence of our shift into the Exponential Age. On the one side, you have the new behaviours, relationships and structures that are enabled by exponentially improving technologies, as well as the products and services built from them. On the other, you have the norms that have evolved or been designed to suit the needs of earlier configurations of technology. The gap leads to extreme tension. In the Exponential Age, this divergence is ongoing – and it is everywhere.

Consider the economy. When an Exponential Age company is able to

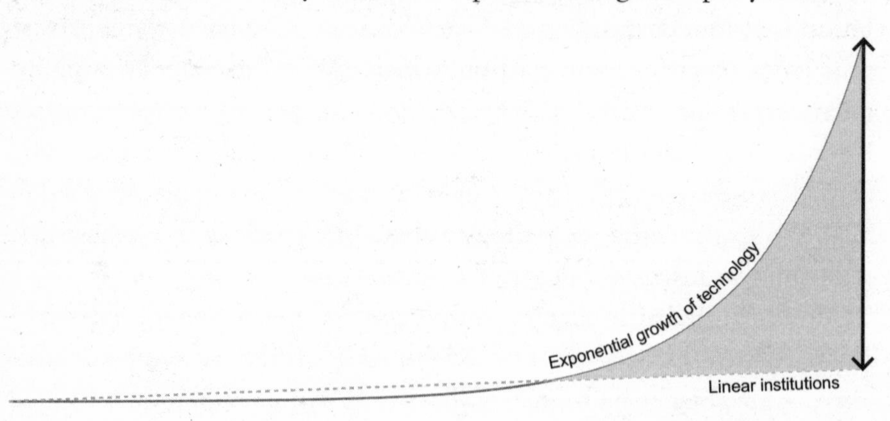

grow to an unprecedented scale and establish huge market power, it may fundamentally undermine the dynamism of a market. Yet industrial-age rules of monopoly may not recognise this behaviour as damaging. This is the exponential gap.

Or take the nature of work. When new technologies allow firms and workers to bid on short-term tasks through gig-working platforms, it creates a vibrant market for small jobs – but potentially at the cost of more secure, dependable employment. When workers compete for work on task-sharing platforms, by bidding via mobile apps, what is their employment status? Are they employees, contractors or something else entirely? What rights do they have? Does this process empower them or dehumanise them? Nobody is quite sure: our approach to work was developed in the nineteenth and twentieth centuries. What can it tell us about semi-automated gig work? This is the exponential gap.

Or look at the relationship between markets and citizens. As companies develop new services using breakthrough technologies, ever-more aspects of our lives will become mediated by private companies. What we once considered to be private to us will increasingly be bought and sold by an Exponential Age company. This creates a dissonance: the systems we have in place to safeguard our privacy are suddenly inadequate; we struggle to come up with a new, more apt set of regulations. This is the exponential gap.

These are the domains that define our everyday existences – as consumers and citizens, workers and bosses. In each area, the potentials of new technologies have challenged the norms of our society. The effect this divergence has will be the emphasis of much of the book. For now, however, our focus is on why the exponential gap occurs.

The most basic cause of the exponential gap is simple: we are bad at maths.

Let's consider for a moment what it actually feels like to live in the

Exponential Age – the answer, as many readers will know, is bewildering. Someone like me, born in the early 1970s, has experienced wave after wave of innovations. From the landline to the mobile, dial-up internet to mobile internet, vinyl records to CDs to MP3s to streaming. I've owned the Bee Gees' *Saturday Night Fever* on at least five different formats: vinyl record, cassette tape, CD, MP3 download, and now, streaming access.

Human cognitive machinery does not naturally process such rapid change. The calculations bewilder us. Take the case of an atypical London rainstorm. Wembley Stadium is England's national soccer venue. It is about five miles north-west of my home, and I see it out of my window when I go to visit my in-laws. Its steel arch, spanning 315 metres and reaching 133 metres at its peak, soars above the silver-grey roof. The venue is an enormous edifice, seating some 90,000 people at capacity.

Imagine sitting in the highest row of level three, the furthest above the pitch you can be – some 40 metres or so above the ground. Rain starts to fall, but you are sheltered by the part of the roof above you. Yet this is no ordinary rain. This is exponential rain. The raindrops are going to gradually increase in frequency, doubling with each passing minute. One drop, then a minute later two drops, then a minute later four drops. By the fourth minute, eight drops. If it takes 30 minutes to get out of your seat and out of the stadium, how soon should you get moving to avoid being drenched?

To be safe, you should start moving by no later than minute 17 – to give yourself 30 minutes to be clear of the stadium. By the 47th minute, the exponential rain will be falling at a rate of 141 trillion drops per minute. Assuming a raindrop is about 4 cubic millimetres, by the 47th minute the deluge would be 600 million litres of water.[7] Of course, the rain in the 48th minute will be twice as large, so you are likely to get soaked in the car park. And if you make it to the car, the deluge in the 50th minute will comprise 5 billion litres of water. It would weigh 5 million tons. Frankly, if exponential rain is forecast, you're best off staying at home.

Exponential processes are counterintuitive. And we struggle to grasp them. Thomas Malthus, the eighteenth-century political economist, first articulated the problem. According to Malthus, the human population tends to grow exponentially – but we won't realise the power of that exponential growth until too late. Eventually, human needs will outgrow our ability to produce food, bringing famine and pestilence. Malthus's dire predictions did not come to pass, thanks to the extraordinary increases in productivity brought about by the industrial revolution. But his basic insight – that we continually underestimate exponential processes – remains pertinent.

As the modern environmental movement began to develop in the late 1960s, it harked back to some of Malthus's warnings. Albert Bartlett, a physicist from the University of Colorado, began lecturing on the limits of population and energy use in September 1969. He observed that 'the greatest shortcoming of the human race is our inability to understand the exponential function'.[8] It was a message he was keen to spread. He presented his lecture 1,742 times over the subsequent 36 years.

Three years after Bartlett began lecturing, the groundbreaking tract *The Limits to Growth* argued that exponential patterns in the rate of resource usage would become fundamentally unsustainable.[9] The book sparked a movement, driven by concern that increasing resource consumption would destroy our biosphere. But less famously, the book also ignited great discussion over whether humans are even able to predict the types of accelerating (or non-linear) changes that our economic activity foments.[10]

Since then, academics from fields as diverse as finance and psychology have uncovered reams of evidence that we are hopeless at understanding exponential growth. Psychologists who study how people save for the future have identified the 'exponential growth bias', which makes us underestimate the future size of something growing at a compounded rate.[11] Research in this area shows how people are consistently befuddled by the compound growth of savings, loans and pension plans. If you started investing in your pension a little late, you – like many of us – may have suffered from a persistent bout of exponential growth bias.

One study tested how well Swedish adults could understand compounding growth processes. Researchers asked an apparently rather muted question: if you put 100 Swedish kronor in a bank account where it would earn 7 per cent interest a year, how much would you have in 30 years? Even that simple growth rate baffled respondents. The median answer was 410. The correct answer, 761 kronor, is almost double that. More than 60 per cent of the respondents underestimated the answer.[12]

And that was people's underestimation of an annual compounding rate of 7 per cent. Imagine our predictions of an exponential technology that improves at 10 per cent or more per annum. Three decades of compounded 7 per cent change results in an increase of a factor of 7.61. A growth rate of around 40 per cent – roughly what Moore's Law describes – would see around a 32,000-fold increase in that time. One peer-reviewed summary from 1975 summarises the issue: underestimation of exponential growth is a 'general effect which [is] not reduced by daily experience with growing processes'.[13]

This blind spot has a close relative – the 'anchoring bias'. The Nobel Prize-winning economists Daniel Kahneman and Amos Tversky have explored how people make decisions amid uncertainty. They found that, when presented with a numerical challenge, people tend to fix upon some readily available number and adjust their responses around it. It's a trick salespeople use: by starting at a particular price, they anchor our expectations about what the real value of something might be. But it fails when it encounters exponential growth. As the growth curve takes off, people's expectations remain anchored around small figures from early in the process.

Why do humans consistently underestimate the power of exponential change? The shortest answer is perhaps the best. Most processes we go through – ageing from one to two to three years old; moving incrementally along a metro line – follow a linear scale. And there are good evolutionary benefits behind this linear mentality. Our minds developed for a world that, in general, hadn't yet discovered the power of rapid change. The rhythms of the hunter-gatherer routine were slow: we adapted for a life that was largely seasonal, based on repeated patterns throughout the year. And this slow pace defined our existence until remarkably recently.

Historians of the pre-industrial world often point to the slow, almost unnoticeable beat at which societies changed. The French historian Fernand Braudel wrote an account of life in medieval France, the first part of which focused on 'a history whose passage is almost imperceptible, that of man in his relationship to the environment, a history in which all change is slow, a history of constant repetition, ever-recurring cycles' – births, weddings, harvests, deaths.[14] This leisurely rhythm of life defined our existence for almost all of human history – even in the most industrialised nations, like the UK and US, it took until the 1920s for half of the population to live in cities. Is it any wonder we have struggled to adapt to a faster pace of life?

There have always been exceptions, of course. The spread of a pandemic pathogen, such as the H1N1 virus that caused the Spanish flu, follows an exponential curve in its growth phase; as does money in periods of hyperinflation, like that of Weimar Germany in the 1920s or, more recently, in Zimbabwe and Venezuela. But apart from those relatively idiosyncratic examples, humans have not often had to deal with dramatic continuous changes within their lifetimes.

We evolved for a linear world. And if a linear world creates one set of experiences, the Exponential Age creates another entirely. As evolutionary psychologists Pascal Boyer and Michael Bang Petersen put it, human minds were likely 'selected for . . . the first kind of adaptive problems, [rather] than the second'.[15]

What happens when we try to make predictions about exponential change in the real world – rather than in psychologists' experiments? I've seen the trouble this poses first-hand. During the course of my career, I've watched well-informed people pooh-pooh mobile phones, the internet, social networks, online shopping and electric vehicles as niche playthings destined for eternal obscurity. Over two decades I've observed executives in established industries regularly, perhaps even deliberately, look at the spread of a new product or service and dismiss it. Often it was because the absolute numbers were small, in spite of signs of hockey stick growth. Like spectators at Wembley Stadium during a period of exponential rain, they didn't leave their seats until it was too late.

For example, in the early 1980s, companies started to operate the first cellular phone services. At the time, handsets were clunky, calls were filled with static, data services were non-existent and coverage was patchy. Yet it was becoming clear that mobile telephony had clear, practical benefits. The giant American phone company AT&T asked the world's top management consultancy firm, McKinsey, to estimate the future market size for this product. McKinsey put together a 20-year forecast: the US mobile phone market would approach 900,000 subscribers by 2000.[16]

Not quite. The first phone, the cement-brick-sized Motorola DynaTAC, cost $3,995 in 1984. But the core components were getting cheaper every year, and phones followed suit: becoming better, smaller and cheaper. By the year 2000, you could find a new cell phone for a couple of hundred dollars. And the capabilities of the networks were growing too. In 1991, mobile networks were just starting to introduce data services – until then phones had been used almost exclusively for voice calls. In those days, if you bought a device to connect your computer to the cell phone network, you could use it to send data at a rate of 9,600 bits (or about 1,000 words) per second. Had digital cameras been widespread at the time (they weren't), sending a single photo would have taken several minutes. By 2020, common 4G phone networks could deliver 30 million bits per second, or more, to handsets.

Mobile tariffs collapsed in line with the growing speed of the networks. Between 2005 and 2014, the average cost of delivering a megabyte of data – the equivalent of about 150,000 words – dropped from $8 to a few cents.[17] In short: McKinsey had miscalled it. In the year 2000, more than 100 million Americans owned a mobile phone. The most storied management consultants in the world had been wrong by a factor of 100. Predicting the future is hard; predicting it against an exponential curve harder still.

Nor is this a problem limited to the private sector. The International Energy Agency is an intergovernmental organisation that was founded in 1974 in the wake of the global oil crisis the previous year. The IEA's annual *World Energy Outlook* has, for years, predicted the amount of electricity generated by solar power. In one of its forecasts, made in 2009, the IEA predicted 5 gigawatts of global solar power by 2015. They were wrong. The actual number in 2009 – yes, the year in which they made that prediction – was 8 gigawatts. In 2010, they upgraded that 2015 forecast to 8 gigawatts. In 2011, they upped it again, to 11 gigawatts. In 2012, they predicted 24 gigawatts. By 2014, they were predicting 35 gigawatts of solar capacity by the next year. The real capacity for 2015? 56 gigawatts. This global group of experts systematically misread the market for six years straight, right up until the year beforehand. But it didn't stop there. After six years of hopeless forecasts, they continued the trend for several more. In 2018, the IEA estimated current global solar capacity was 90 gigawatts. And they predicted a rough standstill for the next year – an estimate of 90 gigawatts for 2019. In reality, 2019's output exceeded 105 gigawatts. For that year, the annual growth they forecast was off by 100 per cent – or infinity, depending on how you do that maths. It was a decade of looking at an exponential technology, dropping in price and increasing in scale, and systematically getting it wrong.[18]

The problem is not just that we underestimate exponential growth, however. Experts who are mindful of the power of exponentiality can also be prone to overestimating its power. In his 1999 book *The Age of Spiritual Machines*, Ray Kurzweil predicted that by 2019 a $1,000 computer would be 'approximately equal to the computational ability of the human brain'.[19] This proved optimistic. When trying to square rapid, exponential growth with an inordinately complex issue, a slight error in your basic assumptions can throw your whole prediction off. And with a neural network as complex as the human brain, it's near impossible to get these assumptions right. Our best current guess is that the human brain has about 100 billion neurons.[20] Each neuron is connected, on average, to a thousand others, leading scientists to estimate that there are 100 trillion connections in the human brain.[21] If these estimates prove correct, and if we have properly understood the function of neurons, a machine that mimics the complexity of the brain could conceivably be built within a couple of decades. But those are big ifs. When our scientific understanding of a subject is still developing, predictions are sometimes little better than guesswork.[22]

Self-driving cars create similar, if rather smaller, headaches. In 2019, Elon Musk envisioned that Tesla, the car firm, would have a fleet of 1 million self-driving taxis, what he called 'robo-taxis', on the roads by the end of 2020.[23] The actual number was zero. And Tesla is not alone. Every self-driving car company has missed its targets. It turns out that the problem is much harder, from a purely technical perspective, than the teams building the technologies were willing to acknowledge. When you jump into the car for a quick trip to the grocery store, you make roughly 160 decisions for every mile you travel. While that might not sound like a lot, when the decisions are based on a near-limitless number of variables, the scale of the challenge comes into sharper focus.[24]

And these problems of underestimation and overestimation are confounded by a third difficulty – unforeseen consequences that don't feature in our predictions at all. Exponentiality often has unexpected effects. Take chewing gum. In the 10 years from 2007, American chewing gum sales fell 15 per cent – just as 220 million American adults bought their first smartphone. This was no coincidence. When people got into a shop queue, they would once have spent the time browsing the goodies for sale at the counter – and gum was the obvious choice. Now they were spending that time playing with their phones. So gum sales plummeted.[25] Nobody saw that one coming. Predicting the impact of the iPhone on grocery-store gum sales would have needed a modern-day Nostradamus.

Our inability to make accurate predictions about exponentiality reached its zenith in 2020. When the Covid-19 pandemic got underway, most of us consistently underestimated how rapidly it would spread. At an early point in a disease's progression, the numbers of new infections may be ignorable. But if the rate of growth is inexorably rising, soon you'll have an issue. In the US, only 60 new Covid cases were discovered in the two weeks from 15 February. The next two weeks witnessed 3,753 new cases, the two weeks after that 109,995. By mid-November, 150,000 cases were being added per day.

In the early days, this exponential growth was something the public and policymakers – in America and Europe, at least – proved unable to grasp. Politicians from Donald Trump to Boris Johnson consistently downplayed the risk exponential growth represented. Early research, released during the first year of the pandemic, demonstrated exponential growth bias at play. At all stages of the pandemic, people underestimated the future course of the spread. Given three weeks of actual data for the growth of the virus, participants were asked to predict infection levels one week and two weeks later. Despite having the data to hand, people guessed the future trajectory of the disease very poorly. On average, participants underestimated the first week by 46 per cent, the second by 66 per cent.[26] Making predictions in the Exponential Age is hard – and we often get it wrong.

If the primary cause of the exponential gap is our failure to predict the cadence of exponential change, the secondary cause is our consequent failure to adapt to it. As the speed of change increases, our society remoulds itself at a much slower pace. Our institutions have an inbuilt tendency towards incrementalism.

The way fast technology runs ahead of our slow institutions is nothing new. This is arguably one of the key, inevitable consequences of innovation. In the nineteenth century, breakthroughs in industrial machinery catapulted the British economy into a position of global dominance. But there was a hitch. There was a 50-year period where British GDP expanded rapidly but workers' wages remained the same – something the economic historian Robert Allen calls 'Engels' pause'. Those with capital to invest in new machinery did well initially, because it was technology that was driving the growth. It took decades for workers' wages to catch up.[27]

The problem was not just wages. The industrial revolution eventually meant more wealth, a longer lifespan, and a better quality of life for all. But for most labourers, the first effect of industrialisation was an – often unwelcome – change in working conditions. Starting in the late eighteenth century, technology moved millions of people from fields, farms and workshops into factories. In the 1760s, before the industrial revolution really took off, the average British worker toiled for 41.5 hours a week. By 1830, this had risen to 53.6 hours – an extra hour and a half each day of week. By the 1870s, when the Victorian economy had largely completed its transition from agriculture to industry, the typical worker was nudging 57 hours a week.[28] And other countries in what is now the developed world didn't escape the demands of early industrialisation. By 1870, the average American was sweating out 62 hours a week; the typical Australian only six hours less.

Perhaps better than any other author or observer, Charles Dickens chronicled how the nineteenth century was a time of exploitation, misery and feculence. In *Hard Times*, published in 1854, Dickens describes the poor living conditions of industrial towns in the north of England, where soot and ash permeates the air outside the factory, and workers get paid next to nothing and yet work constantly. But even before Dickens's time, the working conditions of the new era were infamous. Masquerading as a Spanish nobleman, in 1814 the British poet Robert Southey wrote a description of working life in Birmingham:

> *I am still giddy, dizzied with the hammering of presses, the clatter of engines, and the whirling of wheels; my head aches with the multiplicity of infernal noises, and my eyes with the light of infernal fires,—I may add, my heart also, at the sight of so many human beings employed in infernal occupations, and looking as if they were never destined for*

anything better . . . The noise of Birmingham is beyond description; the hammers seem never to be at rest. The filth is sickening: filthy as some of our own old towns may be, their dirt is inoffensive; it lies in idle heaps which annoy none bar those who walk within the little reach of their effluvia. But here it is active and moving, a living principle of mischief, which fills the whole atmosphere, and penetrates every where.[29]

In his seminal study *The Condition of the Working Class in England*, written 30 years later, Friedrich Engels describes urban living conditions as a 'mass of wretchedness [of which we] ought to be ashamed'.[30] Over 150 years later, Engels would give his name to the eponymous 'pause' thanks to accounts like these.

One way to make sense of the social problems brought by industrialisation is as a gap – between the speed of technological and social change and the speed of institutional and political adaptation. The state's failure to regulate working practices during the industrial revolution reflected the preoccupations of an agrarian and aristocratic elite; Britain had a modern economy, but a distinctly pre-modern political order. As the former British prime minister Tony Blair told me: 'There was a time lag between the change and the policymakers catching up with it.'[31] Factories only started to be properly regulated with the 1833 Factory Act; the emergence of a labour movement and welfare system to look after the interests of the urban working class took much longer.

Like those of our Victorian forebears, today's institutions face the conundrum of keeping up with rapidly changing technologies. But this time, the gap will grow bigger and more quickly. In the Exponential Age, radical change takes place not over decades but over years – or sometimes months.

It's worth pausing to reflect on what these 'institutions' actually are. The word conjures up a sense of solidity: an imposing police building, a large church, or the towering UN headquarters in New York. Yet institutions are not buildings. They are more than that. They are the systems that govern our everyday lives, our individual actions in public and in private, and how we relate to each other.

In sociological terms, institutions are all the lasting norms that define how we live. For our purposes, I consider an institution to be any kind of arrangement between groups of actors in a society that helps them relate to each other. These institutions are what give stability to the coming-and-goings that make up our lives. As the sociologist Anthony Giddens puts it, 'Institutions by definition are the more enduring features of social life.'[32] If they weren't, they wouldn't be institutions – they would be fads, ephemera, crazes.

But within this wide umbrella, institutions can take many different forms. Some institutions might be obviously 'institutional' in nature: a business is an arrangement between employees, bosses and owners; a state between its citizens and the machinery of government. The branches of government are formal institutions. So are religious entities, like churches, mosques, temples or synagogues. As are international organisations: the European Union, the World Bank, the World Health Organization, or the International Energy Agency, the group responsible for those hapless forecasts around solar power. Many such institutions are 'embodied': we can picture the people who make them up and the buildings in which they are located.

Other institutions, like the notion of the Rule of Law, or the body of international agreements and domestic legislation that make up intellectual property law, are not groups of people but are institutional nonetheless. Some may look more like a set of formal rules than an organisation. Laws, and the legal systems in which they exist, are often written down – these are one of the primary institutions governing our lives.

But not all institutions need to be so formal. There are the habits and practices that guide our behaviour. Such unwritten rules can have as high, perhaps higher, levels of adherence than written ones. For many people who play the board game Monopoly, fines are placed in the centre of the board, to be grabbed as a bonanza by anyone landing on 'Free Parking' – even though this isn't in the rulebook. In the UK, it is common practice to flash one's headlights to thank a driver who gives way to you, even though the official Highway Code advises against it. In an example used by the economist Richard Nelson, bakers have a shared best practice for making a cake, even if there might be differences in the details.[33] These norms are also a type of institution. They bring stability that frames our collective behaviour.

All these institutions have something in common. They are largely not cut out to develop at an exponential pace or in the face of rapid societal change. In the most extreme cases, they're not cut out to adapt at all.

Take one of the most institutional institutions in history: the Catholic Church. Nearly 2,000 years old, the Church is one of the longest-standing organisations in existence. In 1633, it got into a dispute with the astronomer Galileo Galilei and his conclusions about the structure of our solar system. Yet it was not until 1979, 346 years after the astronomer was condemned to house arrest until his death – and 22 years after the Sputnik satellite orbited the planet – that Pope John Paul II ordered a papal commission to review Galileo's conviction.[34] Some 13 years later, the commission and the pope overturned the findings of the seventeenth-century inquisition that denounced Galileo's subversive ideas.[35]

This is an extreme example. But even if few institutions are as slow to adapt as the Catholic Church, almost none are particularly fast-moving. This is the inverse of the difficulty of predicting exponential change: even when our predictions are correct, our institutional responses can still be lackadaisical. Take the Kodak company. In 1975, a Kodak engineer called Steve Sasson put together a device, about the size of a toaster, that could save images electronically. A 23-second process transferred the images to a tape, where they could be viewed on a TV screen. At the time it was astonishing; personal computers barely existed as a category and video recorders were a rarity in American homes. A couple of years later, Sasson was awarded US Patent 4131919A for an 'electronic still camera'.[36] He figured, by extrapolating from Moore's Law, that it would take 15–20 years for digital cameras to start to compete with film. His estimate was bang on. But Kodak, even with a two-decade head start, did not grab the opportunity. At the time, Kodak sold 90 per cent of the photographic film in the US and 85 per cent of its cameras – they felt little need to completely pivot their business strategy. 'When you're talking to a bunch of corporate guys about 18 to 20 years in the future, when none of those guys will still be in the company, they don't get too excited about it,' Sasson later recalled.[37]

Kodak did go on to develop a range of digital cameras; in fact, theirs was one of the first to market. The company even recognised the power of the internet, buying Ofoto, a photo-sharing site in 2001, nine years before Instagram was founded. But institutional knowledge – the established consensus about what their business was about – held them back from building anything nearly as popular as Instagram, which would become the most successful photo business in the world. Kodak's execs viewed Ofoto as an opportunity to sell more physical prints (their old business), not as a chance to connect people via photos of their shared experiences (what the internet would offer). Indeed, the rationale to hamstring Ofoto by linking the site to Kodak's traditional film-and-prints business was the same wrong-headed decision the firm had taken more than two decades earlier with the digital camera. The market for film cameras had peaked in the late nineties – the world had moved on, propelled by technology. Kodak had not: the institutional memory was strong. Eventually, the firm hit the buffers, struggling and eventually folding in 2012. Kodak sold Ofoto as part of its bankruptcy process.[38]

Nor are more technologically minded organisations immune to such slowness. Microsoft is one of the most dynamic firms in history, yet it too has been blindsided by the pace of change. In the mid-1990s, Microsoft was slow to spot the disruptive power of the internet. It was not until 26 May 1995 that Bill Gates, the then CEO, wrote the now-famous memo where he argued: 'The Internet is a tidal wave. It changes the rules. It is an incredible opportunity as well as [an] incredible challenge.'[39]

It was already too late. For a couple of decades, Microsoft would remain a smaller player in the internet domain – at least compared to its dominant position in computer software. And it ceded key parts of the internet business, be it the e-commerce or internet search or messaging markets, to other companies. Faced with the launch of the iPhone in 2007, the firm flubbed again. Steve Ballmer, the then CEO of Microsoft, dismissed Apple's new gadget, saying: 'It has no chance of gaining significant market share.'[40] Of course, the iPhone went on to become the dominant mobile device in the United States and Microsoft's own mobile ambitions were left in tatters. The company launched its first Windows software for mobile phones in 2000 only to mothball the project 15 years later, with the market all but left to Apple's iOS and Google's Android software.

Ballmer puts his mistake down to applying his strong understanding of the PC industry, the previous paradigm, to the new smartphone industry. In a sense, he fell into the exponential gap. He stepped down from his role in 2014, after it became clear that Microsoft had miscalculated the prospects of Apple's iPhone.[41] (The company would not miss a third time. Satya Nadella, the CEO who replaced Ballmer, would successfully pivot the company to take advantage of yet another technology shift in the 2010s – the rise of cloud computing for businesses – to help Microsoft reclaim its position as one of the world's most valuable and innovative companies.)

It's not just companies that come unstuck in the face of rapid change, of course. Our wider legal and political institutions can prove intransigent too. The sociologist William Ogburn, writing in 1922, described a 'cultural lag' in the wake of new technologies like industrial machinery and the car. His conception of culture was similar to my description of institutions – a broad category that encompasses a wide array of habits and norms. In one of his case studies, he examined the rate of growth of industrial accidents and compared it to the rate at which various American states developed compensation and liability statutes. The first of these laws was passed in 1910; nothing was enforced nationally for a further 12 years. Ogburn suggests that, even 40 years before the first employer's liability law was passed, workplace accidents were clearly a problem: '1870 was hardly too soon to have developed workmen's compensation acts. Had they been in force in the United States from 1870 on, a very great many accidents would have been cared for with much less burden to the worker.'[42]

More recently, the rise of new technologies has challenged the woollier cultural conventions of our daily lives. Is it reasonable to use a smartphone while at dinner? Teenagers may think so; their parents may beg to differ. The smartphone is a transformative cultural force. It is not just a teenage phase – it establishes a whole new set of behaviours and values, to which many groups may struggle to adapt. Ditto video games. Gaming, especially online gaming, has become an increasingly critical element in friendship and socialisation among teenage boys.[43] But our cultural norms have struggled to keep up. A decade and a half after the arrival of the smartphone, and two and a half decades after the first PlayStation, the rules remain unclear.

Why are institutions – whether nebulous social norms or blue-chip companies – so slow to change? Many studies on this subject point to 'path dependence', the idea that practices and customs become relatively determined from an early point in any process.[44] Sociologists point out that the path an institution starts out on can have long-running effects on our behaviour – we get 'locked' into a particular course of action. Let's return to the smartphones-at-dinner conundrum. The norm of shared family dinners in America developed in the eighteenth century, among wealthy families who had the money to build a separate room for dining – featuring a dedicated table at its centre.[45] Over time, the cultural importance of the dining room evolved – leading to the establishment of set mealtimes, and moral judgements about families who did not sit down to eat together. You can chart a direct course between these eighteenth-century origins and today's popular parenting wisdom regarding the importance of family dinners – from an early point, that cultural path was set.

This phenomenon has wide-ranging implications for all manner of institutions – ranging from businesses to governments to social norms. Even when we do try to change an institution, it tends to remain rooted – directly or indirectly – in what has come before.

The leading theorist of institutional change, Kathleen Thelen, has identified a number of key ways in which institutions adapt in practice. These include layering (when new norms get built on old ones), drift (where an institution keeps its policies in place, in spite of a changing context) and conversion (when an institution takes an old way of doing things and redeploys it in a different context). But all hint at the fact that institutions get locked into outmoded ways of responding to the issues they face.[46]

A good example of layering comes from the UK's National Health Service, which until 2020 was still regularly using pagers to exchange messages. The NHS started using them in the 1980s, before the age of cheap mobile phones. The more they were used, the more embedded they became in NHS systems. And the more embedded they were, the more practices and norms were built on top of them – even long past the point when they had been surpassed. New institutions built on the back of pre-existing ones. Layering in action. For my part, I personally experienced institutional drift when helping the BBC, the publicly funded UK broadcaster, launch some of its first websites in the late 1990s. The television execs who were handed the brief took their cues from television – shows had transmission dates and episodic release cycles (such as once a week) – and brought them to the world of real-time publishing on the web. It was an uneasy match, and did not sit well with the expectations of web surfers.

There are exceptions to this institutional slowness, of course. On occasion, institutions can lend themselves to very rapid change. Wars and revolutions help. It took less than a year to create the International Monetary Fund after it was proposed at the Bretton Woods Conference in July 1944. The visceral shock of World War Two and the need to find a solid base for international cooperation provided the impetus to establish many institutions, such as the United Nations and the General Agreement on Tariffs and Trade.

These are, in the language of institutional theory, moments of 'punctuated equilibrium'. But on one level, this caveat merely reveals the sheer scale of the disaster needed to shock institutions into rapid change. In the absence of a catastrophe, institutions tend to develop more like the Catholic Church than the frantic establishment of the United Nations.

Put together these two forces – the inherent difficulty of making predictions in the Exponential Age, and the inherent slowness of institutional change – and you have the makings of the exponential gap. As technology takes off, our businesses, governments and social norms remain almost static. Our society cannot keep up.

In the early years of the twenty-first century, this exponential gap was relatively trivial. The demise of Kodak or the failure of Microsoft to understand the potential of the early internet were not society-threatening problems. In the early years of the computer industry, technology was an isolated, niche industry. In those early days, if a company failed to adapt to a new trend the consequences were fairly mild. Firms might go bust, leaving customers nursing defunct products and former staff with fond (or not so fond) memories. But that was not the end of the world.

However, after we tipped into the Exponential Age, the gap became a more existential problem. As of the early 2020s, exponential technology has become systematically important. Every service we access, whether in the richest country or the poorest, is likely to be mediated through a smartphone. Every interaction with a company or government is (or soon will be) handled by software powered by a machine-learning algorithm. Our education and healthcare is increasingly delivered through AI-enabled technologies. Our manufactured products, be they household conveniences or houses, will soon be produced by 3D printers. Exponential technologies are becoming the medium through which we interact with each other, the state and the economy.

For the people and companies who understand this shift, the exponential gap creates a huge opportunity. Those who harness the power of exponentiality will do much better than those who don't. This isn't simply about personal wealth. Our rules and norms are shaped by the technologies of the time – those who design essential technologies get a chance to shape how we all live. And these people are in the minority. We are witnessing the emergence of a two-tier society – between those who have harnessed the power of new technology, and those who haven't.

What is to be done? When we return to our diagram of the exponential gap, it becomes clear that there are two potential ways to stop our social institutions being outpaced.

On the one hand, we can close the gap by slowing the rise of the top line. But this is no mean feat. For one thing, this acceleration is already baked into the structure of our economies. The process through which technologies improve and accelerate is not centrally controlled. It emerges from the needs of individual firms, and is met by a coalition of players across the economy. The virologist benefits from a faster genome sequence, and so seeks out better electrochemistry, faster processors, and more accessible storage for genomic data. The householder wants

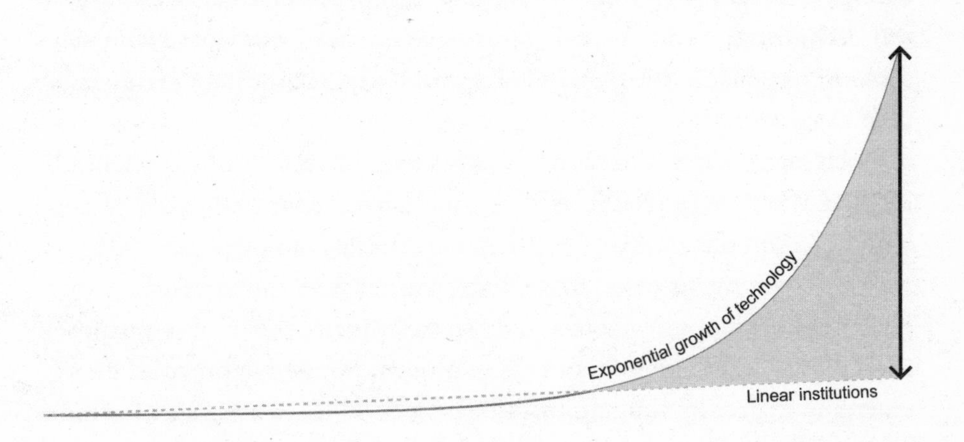

more efficient solar cells; the farmer more precision methods to fertilise her crops. The Exponential Age is a near-inevitable consequence of human ambition.

Even if we could slow the pace of exponential change, it's not clear this would be desirable. Many of the most urgent issues of our time can only be solved with exponential technology. Tackling climate change, for example, requires more radical tech, not less. In order to decarbonise our economies, we will need to rapidly shift to renewable sources of energy, develop alternatives to animal proteins for food, and scale building materials that have a zero-carbon footprint. What's more, figuring out how to deliver good-quality healthcare, education, sanitation and power to the poorest billions on the planet is another problem that technological innovations can address. The expensive (and resource-intensive) way that the developed world achieved those outcomes in the past is not practical for poor countries at a time of environmental crisis.

So, putting the brakes on the development of technologies is hard to justify. A preferable option is to better prepare for a period of accelerating technologies. We can close the gap by making the lower line rise faster. That means equipping our social institutions – our governments, companies, cultural norms – to adapt at pace. At the same time, we can try to make our institutions more prepared for rapid change – so they aren't undone by any bumps in the road of exponential technology. And we can even develop new ways of organising society that ensure the benefits of exponential technology are distributed evenly.

Such measures would allow us to harness the power of exponentiality, and the rules and norms that can shape it, for the needs of our societies. And this is what I will advocate in the coming pages.

This is an urgent need. In the Exponential Age, the institutions that govern our economies today will soon cease to be fit for purpose. Over the next five chapters, we'll explore how the exponential gap is emerging in sectors ranging from business and the workplace, to trade and geopolitics, to the relationship between citizen and market. In each, new technologies will clash with our existing expectations, rules and systems. And in each, we will need radical thinking to prevent the exponential gap from eroding the fabric of our society.

4

The Unlimited Company

For the first three decades of my life, there was a set of clear rules about how business worked. Dozens of very large companies dominated the economy, and each had a specialty. Exxon and BP invested in oil rigs and drew oil from the ground, selling it at a market price that was above the price of extraction. General Motors and Ford assembled raw materials and components, stitched them together and sold them for a profit. Successful companies grew bigger, by dint of better products or keener pricing. They might benefit from economies of scale – as they got bigger, their costs, relatively speaking, went down.

But these companies also encountered forces that held them back. As they became bigger, they might become too complex to manage. The flows of information between bosses, managers and employees might get entangled. Headquarters might lose track of what the front line worker was doing. If you have ever worked at a very large company, you'll be familiar with how bureaucracy creeps in and slows everything down. The economist Ronald Coase described this predicament well: in his rendering, organisational costs grow as firms get bigger and come to act as a force of gravity, slowing a company's expansion and eating away at the advantages of scale.[1]

As companies expanded and became more complex, managing them became like sorting through a mess of tangled wires – with too many contradictions resulting in slower and slower decision-making. To escape this problem, firms might instead develop a narrow specialty. By focusing on just one thing, a company would carve out a competitive advantage – but it would also become less entrepreneurial, and less capable of seizing new opportunities in adjacent sectors. The organisational muscle and the institutional flex they honed would emphasise execution and efficiency, not exploration and agility. According to the complexity scientist Geoffrey West, as companies got larger, they tended to become untenably complicated. Big and old firms became 'short-sighted, conservative and unidimensional', he says.[2]

This resulted in a counter-intuitive phenomenon – for established, large firms, it got increasingly expensive to get bigger. As size conferred both strength and weakness, competitors might be able to chip away at the advantage of the market leader. Dominance, in the twentieth century, was hard.

In industry after industry, one could feel the gravitational pull of scale. It led to a phenomenon called diminishing marginal returns – that is, a decreasing return for every dollar a company invested. Companies would normally get to a certain share of a market and no bigger – two-fifths was about as good as it got. In the pharmaceutical industry, the top firms might have 40 per cent market share.[3] In groceries, Tesco (the top food retailer in the UK) had about a 25 per cent market share in 2019. In the traditional car industry, the market was extremely fragmented – in the UK in 2020, no manufacturer held more than a 12 per cent share.[4] In televisions, Samsung led the global market with an 18 per cent share.[5] Even the gigantic Coca-Cola company, whose market share in the US did exceed 40 per cent, was not even double the size of PepsiCo, its closest competitor.

This level of competition was meant to keep these companies honest. If they tried to raise prices, a competitor could step in. In the odd instance where a single firm did have a big market share – 50 per cent, 60 per cent or more – everyone acknowledged you had a problem. You might find bad behaviour, like price gouging to force out competitors – invariably followed by price rises once the company's position was secure.

Such market dominance might even provoke the ire of the state. The standard story is Standard Oil. At one point, John Rockefeller's firm controlled 90 per cent of the US's refined oil. It became so dominant that the US Supreme Court ruled it an illegal monopoly, and the firm was splintered into 34 not-so-little pieces in 1911.[6]

Yet when we turn to the corporate titans of the Exponential Age, we see a very different picture. Google's market share of search queries is almost 80 per cent in the US, 85 per cent in the UK, and nearly 95 per cent in Brazil. In the smartphone market, Android is installed on four out of five phones globally, with Apple's iOS used on the vast majority of the others. No other operating system really figures. This dominance is even more pronounced among certain demographics and age groups. In the US, more than 85 per cent of teens own an iPhone.[7] In online advertising, Facebook and Google between them account for more than 90 per cent of all global spend.[8]

The pattern holds in more traditional markets. In the world of cab riding – now given the flashy name of 'ride sharing' – Uber controlled 71 per cent of the US market in 2020.[9] There's a similar picture in retail, historically a cut-throat and highly competitive market. America's market-leading retailer, Walmart, had a little over a quarter of the US's offline retail market in 2017. Yet when it comes to the internet, Amazon's market share across online commerce looks set to grow well above its current level of around 40 per cent – a number that is already two-thirds higher than Walmart's offline share.

Exponential technologies seem to imbue companies with powers that allow them to defy the force of gravity that held back firms of earlier generations. Economists call this new type of firm the 'superstar company'. The superstar company rises rapidly, seemingly unburdened by the forces that hold traditional firms back. It seems to be more productive, more aggressive, more innovative, and able to grow faster. It dominates markets that already exist and creates markets that didn't exist before. Superstar firms get bigger and bigger, dominating one market, then the next.[10] Many superstar firms are household names – Apple, Google, Uber, Facebook. They transform the markets they inhabit, turning them into fertile ground for themselves and parched deserts for their competitors.

Superstardom is becoming commonplace around the world – in China, the United States, across Europe and beyond. The management consultancy McKinsey noted in 2018 that superstars 'come from all regions and sectors'.[11] By 2015, the top 50 firms in the world comprised nearly 50 per cent of the sales of the top 500. That proportion was only around 42 per cent in 1975.[12] About 10 per cent of the world's public companies create four-fifths of all company profits.[13] The more digitised an industry – in other words, the more it relies on the logic of Moore's Law – the more prone it seems to superstarification. The top 25 per cent of IT firms were four times more productive than the bottom 25 per cent in 2015. But that gap was only a factor of 1.5 in the shoe and cement industries.[14]

Superstar firms have broken the corporate culture of traditional markets. The large firms of yore – the Toyotas, Walmarts, Santanders – fought for every point of market share. Their managers delighted in small gains. Winning half a per cent of a market could mean a huge payday for the executives who delivered it. Not any more. In the exponential economy, the winner takes all. While we've always had firms that do better than others, the difference between the best and the worst is greater now than ever.

In this chapter, we'll explore why today's largest firms tend to superstardom. It is rooted in technology, and the peculiar ways that exponential technologies behave differently to industrial-era technologies. The rapidly declining costs of exponential technologies are part of the cause. Another, as we will soon learn, is a positive feedback loop that turns market leaders into a kind of perpetual-motion machine. The consequence is that Exponential Age firms achieve market dominance almost by dint of operating in any given market. They don't need to turn to the monopolists' tactics of yore – cornering the market and raising prices, for example. Massive scale emerges organically.

And while superstardom may be beneficial for shareholders in these firms, it is by no means clear that it is good for the economy, or for society. We have long known that for all its productive vigour, capitalism can bring out some of the worst attributes of humanity. It is why we have bodies of corporate law that govern how firms should operate, and that define the obligations of the directors who steward them. It is why many products, such as the loans and credit cards sold by banks, or the drugs pharmaceutical firms launch, are closely regulated. It is why there is a comprehensive set of rules that explains how companies should account for their operations – and ultimately pay tax.

In the age of superstar companies, this system is in tatters. There is an exponential gap – a mismatch between the ways superstar firms operate, and the norms, conventions and rules used to keep them in check. The rise of these companies raises a raft of questions that make our existing norms and rules seem antiquated – from the appropriate response to uncompetitive markets, to how to tax unprecedentedly large and global companies. The exponential gap is taking over our companies, and in turn our economies.

Why does this gap emerge between superstar firms and the rest? In the new economic landscape, any small advantage can be turned into a bigger one and then, possibly, a lasting one. On the flip side, any small slip-up can cause a company to cascade into irrelevance. But why?

It is possible to discern three forces causing the emergence of superstar companies, each of which drives – and reinforces – the other two. The first lies in one of the drivers of exponentiality that we met in Chapter 2: the emergence of new networks. As we've seen, all exponential technologies are the result of a slew of new forces that emerged in the global economy after about 1970. However, the benefits of these forces – in this case, new global networks of information and trade – have not been evenly distributed. They have boosted the rise of superstar companies, thanks to a phenomenon known as the 'network effect'.

I remember the day in 1986, during my summer vacation, that my parents – who by then ran a small accountancy practice in the east end of London – first got a fax machine. Like most office technology of the day, it was beige, noisy and expensive. It didn't get much use in the first few months, but by 1987 it was humming regularly. It soaked up more heat-sensitive paper by the day. More of my parents' clients had bought fax machines, and they increasingly sent information about their accounts by fax rather than in the post. Each additional client was another person my parents could send documents to electronically. Every month, another client or two bought a fax machine: they weren't just faxing my parents, but also suppliers and customers – who were themselves increasingly buying fax machines. With every extra recipient, our fax machine became more useful. When only one person has a fax machine it is effectively useless; when thousands have one, suddenly it comes in handy.

In other words, the benefits the fax machine brought in 1988 – when my parents knew of hundreds of other companies with fax machines – were far larger than the benefit it had provided when they bought it two years earlier. Economists call this sort of additional advantage an 'externality' – that is, an outcome that is external to the buyer and the seller, instead affecting a wider group who aren't involved directly in the transaction.

The first people to write about externalities, the late-nineteenth-century economist Alfred Marshall and his protégé Arthur Cecil Pigou, pointed out that externalities can be positive or negative. Pigou noted that air pollution – caused by the buyer and seller of a car, but affecting everyone – is a negative externality. With my parents' fax machine, we were encountering a positive externality: the 'network externality'. In other words, the benefits of the device went far beyond just my parents' office in east London, increasing the usefulness of the fax machines of everyone else in the network.

This network effect – in which the addition of every new member of the network increases the value of the network for everyone – is not a new phenomenon. It is why, since antiquity, people have gathered in physical markets: you really do benefit from being where all the buyers and sellers are. Every new buyer makes it more attractive to a seller to be there, at least in theory.

Such network effects are one reason why many Exponential Age firms, like Facebook, PayPal, Microsoft, Google and eBay, are so large and successful. As we've seen, these companies were made possible by the exponentially increasing power of computing. But the network effect is what drives them to ever-bigger gains. Most people who use a consumer social network use Facebook because that is where everyone else is. Businesses accept Visa or Mastercard because they are the platforms with the most buyers on them. This positive feedback loop makes the business better and stronger as it gets bigger.

Take the growth of Microsoft. The company has dominated the operating system market for personal computers since early in the history of computing. From the 1980s, slowly but surely, Microsoft secured more and more market share, even as well-funded competitors like IBM and Novell released rival products during the 1990s that might have been technically superior. By the start of the 2000s, Microsoft had a 90 per cent market share of the personal computer market. Even in recent years, it has never fallen below 75 per cent.

Why has Microsoft been so successful? Thanks to the power of the network effect. People writing the software that makes computers useful needed to decide which operating system to support. Each operating system would require separate (and expensive) programming efforts. During the 1990s, there were a number of rival operating systems to Microsoft's DOS (which would become Windows). But software developers wanted to reach the largest market. The consequence was that small advantages mattered. Whichever operating system eked out a tiny lead in market share would take priority amongst programmers. Once Microsoft took pole position, its operating system became the dominant choice for developers. This meant more software was available on DOS (and then Windows) than on rival systems. In turn, this made Microsoft's operating system more desirable for users: they had a wider choice of software and hardware to choose from. This gave Microsoft a larger user base, again making it more attractive to software developers – and so the cycle continued.

Such network effects make it harder, though not impossible, for competitors to break into a market. In the 1980s and 1990s, Microsoft won not only the operating system war, but also the productivity software war. It competed with – and beat – several companies producing everyday office tools. Lotus Development Corporation was founded in 1982 by Mitch Kapor, the man who, three years earlier, had invented the first spreadsheet, VisiCalc. Lotus produced more spreadsheet applications, like 1-2-3, which was the dominant office staple through the 1980s and much of the 1990s. They also made an exotic spreadsheet program called Improv, which I was partial to, as well as Ami Pro, the word processor on which I typed up my undergraduate thesis on the Cuban missile crisis. And Lotus was not alone. Firms like WordPerfect and WordStar also created word-processing software.

Some of these companies were dominant until relatively late: WordPerfect was the market leader in word processing in 1995, when it had 50 per cent market share. But once Microsoft had established dominance in the operating system market, it became increasingly hard to take on. There was a powerful network effect: once everybody else was using Windows, and exchanging Word documents and Excel spreadsheets, it became much easier for you to use them too. The network effect driving Microsoft's success allowed it to spread inexorably from one market area to the next – first operating systems, then word processors, then spreadsheets.

Network effects aren't just the preserve of profit-seeking firms. Wikipedia, the free online encyclopedia managed by a foundation, also benefits from network externalities. Anyone can create competitors to Wikipedia, and there are many niche ones (such as Wookieepedia, dedicated to the *Star Wars* universe). But Wikipedia is where the majority of contributors gather to share their wisdom and critique each other, so it is where readers congregate. And because this is where the readers gather, Wikipedia becomes the space where contributors want to write.

It is a positive externality: every additional contributor brings in more readers, which in turn attracts more contributors.

In fact, the World Wide Web itself benefited from the network effect. When Tim Berners-Lee first developed the web in 1989, there were several contenders for methods of storing information on the internet, like Gopher and WAIS. There were also commercial products like GEnie, CompuServe and Delphi. In 1994, Microsoft even targeted the market with its own offering, Microsoft Network. But Berners-Lee's web became not just dominant but the sole viable information network on the internet, because of network externalities – once a lot of people were there, everyone else had to congregate there too.

Wherever there are network effects, there is the chance of a winner-takes-all market. Once a company has established itself as the market leader, it becomes extremely difficult to challenge it. And this effect is being driven by the consumers themselves. It is in the consumers' own interest to join the biggest network – it is there, after all, that they will get the most value.

It might not be obvious why these network effects become more pronounced in the Exponential Age. After all, network effects explain much of the pre-exponential world. They were well understood by Theodore Vail, the president of the Bell Telephone Company (the precursor to AT&T), who wrote in 1908: 'A telephone['s] value depends on the connection with the other telephone and increases with the number of connections.'[15]

But the digital technologies of the twenty-first century are better placed to take advantage of network effects than earlier innovations. Why? Because where earlier companies had to go through the cumbersome process of supplying every new customer with a phone, or a fax machine, or some other device, today's digital giants can rely on their customers already accessing a communication network: the internet. The connections that support the network effect are already there.

When I backed an online estate agent in France in 2000, we found that not only did we have to buy internet connections for the realtors, in many cases we needed to supply them with personal computers. Today, anyone – a realtor, an online matchmaker, a rentals business – can take it for granted that most potential consumers will be connected to the internet via a smartphone.

This has created room for a new type of business. Rather than being suppliers of goods, firms can think of themselves as 'platforms' – meaning they connect producers to consumers, without themselves doing much producing (or consuming).

We can see the scale of this shift by looking at an industrial-age company like Ford. The firm organised itself according to its internal processes. Raw materials came in and might be processed in several sequential steps, until there was a completed product – a car. This would be passed out of the business over to marketing and sales executives, who would get the product into the hands of customers – through advertising, well-run car showrooms, and so on. This step-by-step process was known as the 'linear value chain'.[16] The organisational structure of such a company, from goods in to manufacturing, packaging, distribution, marketing and sales, reflected its internal value chain – designed around a series of sequential steps.

Now compare that with a company like eBay. Rather than being rooted in a production line – a linear value chain – it is a platform. Companies like this acquire, match and connect different customers to each other.[17] And so eBay is not a shop. It is a space where sellers can meet buyers to sell their own goods, with eBay itself producing and selling nothing. Facebook is similar. What does Facebook actually offer the consumer? The chance to connect with other people.

This type of offering – a service that matches different types of people – is not really that alien to us. After all, this is what a market does. Marketplaces bring buyers and sellers together to get them to transact. But physical marketplaces have always faced physical limits. They can get too big or too crowded. Anyone wandering the maze of the

Grand Bazaar of Istanbul, or sweating in the alleyways of Anarkali in Lahore, or padding with tired legs around one of Westfield's air-conditioned mega-malls in Australia can attest to this.

And this is why digital technology has led to the emergence of massive new platforms. These all have, to a greater or lesser degree, the quality of the marketplace, bringing together different groups and waiting for magic to happen. But they have no real limitations of scale. The rise of one of the first general purpose technologies of the Exponential Age – the internet – has allowed us to connect far larger numbers of buyers and sellers, without any rooms getting unbearably full or anyone getting tired legs. This unleashed power of platforms is the second key driver of the rise of superstar companies.

Unburdened from the sweaty, crowded, noisy atmosphere of large physical marketplaces, these companies can grow to extraordinary sizes: eBay brings together 185 million active buyers per year;[18] Alibaba, 779 million users.[19] By the summer of 2020, 50 million Americans were using the video service TikTok daily to find their next funny video.[20] And such platforms allow a greater range of choices than any physical marketplace could offer. Take GOAT, the specialist sneaker marketplace. At the time of writing, it's showing me 7,434 size 9 second-hand designer sneakers. The most expensive are an absurd $30,000 for a pair of 3/8 Retro Pack 'Kobe Bryant' Air Jordans. At the other end of the spectrum, a pair of old-school Vans go for a thousandth of that price. As a buyer, there is a cornucopia of choice. And so I'm more likely to visit GOAT in the future. Imagine how big and smelly a physical bazaar that offered 7,000 different pairs of used sneakers would need to be.

The growth potential of these companies is turbocharged further by another factor: platforms are incredibly capital-efficient. The platforms facilitate an exchange without actually having to spend any (or much) money. The world's largest cab company, Uber, owns no cabs and employs no drivers. Airbnb hosts more overnight guests than any hotel chain, yet owns no hotels. Alibaba is the world's largest online showroom for businesses, yet holds no stock.

The marketplace takes care of much of the business's actual logistical processes. Whereas the executive of yesteryear had to worry about where they would find products, when to buy the warehouses to store them, and how to ensure deliveries happened on time, the modern platform executive has done something smarter. Instead, they rely on third parties using their platform to provide stock, warehouse and logistics themselves. The question about who pays for all this has been answered – in the favour of the platform. It's not the platform operator but the wider 'ecosystem' that has to worry about where customers are coming from, whether there are enough vehicles, if there is enough stock.[21]

All this means that the platform business model is increasingly ubiquitous. Exponential Age platforms can creep into any geography and any industry. The model has taken hold with Gojek in Indonesia and Jio in India, as well as Jumia in Nigeria and Bol.com in the Netherlands. And we'd be wrong to associate platform models with lower-value business activities like co-ordinating deliveries or taxi rides. The online medical consultation business Ping An Good Doctor was spun up in 2014 as a service of a well-established Chinese insurance conglomerate, Ping An Insurance. To cultivate their digital medicine platform, they hired 1,000 doctors who helped train AI algorithms. As the algorithms got better the service scaled, allowing AI to respond to 75 per cent of the 670,000 consultations each day. A team of 5,000 specialist doctors digitally handles more complex questions. None of those 5,000 specialists are employed by Ping An; they jump on the service as needed, much as an Uber driver clocks on to find a ride. And these are just a few examples. In the Exponential Age, platforms are everywhere.

From this heady combination of network effects and platforms, it becomes possible to make out the contours of the exponential corporation. The interaction of these two forces leads some organisations to rapidly become dominant – and other, smaller, perhaps more traditional companies to get left behind. But there's a third force that makes these platform-based, network-driven companies even more unassailable – their basis in the 'intangible economy'.

The economy of the machine age was dominated by things you could stub your toe on. Cars, washing machines, telephones, railway lines – for centuries, the most important products were built in the physical world. And the companies that supplied these products were the corporate titans of their time. You could work out a company's value by focusing on the land and buildings it owned, the machines in its factories, and the stocks of products waiting to be sold. This physical stuff is known as a company's 'tangible assets'.

In the Exponential Age, value is rather more nebulous. Many of today's firms do not get their value from the physical products they move around. That is not to say that during the Exponential Age companies won't build physical things. Hugely expensive, customised factories to build semiconductors are clustering in high-tech areas; shallow seas with stable, strong wind patterns are being populated with grids of titanic windmills. But the true source of these companies' profits is more ethereal. The software that runs Google's search engine, the data that represents the network of friends in Facebook, the designs and brand identity of Apple, the algorithms that recommend your evening's viewing on Netflix – this is where their true value lies. Economists call these non-physical assets 'intangible'.[22]

The speed of the shift to intangible assets has been remarkable. In 1975, about 83 per cent of the market value of companies on the Standard and Poor 500 stock market index comprised tangible assets. Intangible assets accounted for the remaining 17 per cent. By 2015, those proportions had reversed. Only 16 per cent of the S&P 500's value could be accounted for by tangible assets, and 84 per cent by intangibles.[23] This ratio skews even further when you look at the largest firms in the world: the exponential superstars. In 2019, the book value of the five biggest Exponential Age firms – Apple, Google, Microsoft, Amazon and Tencent – ran to about $172 billion. But this measure emphasises a firm's tangible assets and largely ignores its intangibles. To figure out a firm's book value you tally up a company's cash, what it is owed by customers, and its physical assets, and then subtract its liabilities. The stock market valued those firms, at the time, at $3.5 trillion. In other words, traditional assets represented only 6 per cent of what the market thought these companies were worth – their intangible assets were doing the rest of the work.[24]

The changing relationship between tangible and intangible assets can be seen in the make-up of the goods all around us. Today, the value of a product tends to come from the early stages of its creation (its research and design) and in the later stages (the branding and service). The actual manufacturing process is becoming less and less important. This is most obvious with high-tech goods. What would your iPhone be without the App Store and the millions of developers creating software for it? But this shift to intangibles is also true for your cup of morning coffee. The cost of the actual beans is decreasingly significant to the value of your cup, especially if you are drinking a capsule coffee or something from a branded coffee shop.[25]

What has caused this migration? The most obvious driver is the increasing complexity of products. Complex products require know-how rather than stuff to put together. A smartphone is light and made up of small amounts of silicon (in the form of glass and the chips inside), a bit of lithium (in its battery) and aluminium (in its case). Some other elements complement it. The thing that makes the smartphone possible is, yes, the raw materials. But it is the know-how that goes into the manufacturing process – turning sand into a chip, and then connecting that chip to thousands of other components in just the right way – that makes it worth so much. Equally important is the growing size of global markets. Intangible knowledge is expensive to produce the first time: think of the expense of product design, or the manuscript of a new book. But it is cheap to copy. This means that intangible goods are more commercially important in bigger markets: you can offset the disproportionate upfront cost thanks to the large number of buyers.

But whatever its cause, the key outcome of our newly intangible economy is to further catalyse the creation of winner-takes-all – or winner-takes-most – markets. In the intangible economy, companies can scale at an unprecedented pace, and to an unprecedented size. Compare a taxi platform like Uber to a cab company that owns its own vehicles. Uber's algorithm – its key intangible asset – is highly scalable, whereas the traditional cab company needs to add a new cab each time it wants to grow.

Software businesses all follow that same principle. The upfront cost of the first copy of the software is enormous. However, every subsequent copy is essentially produced for free. If the network effect and the platform are the foundations of the exponential company, then intangible assets are the bricks that make it up.

The tendency of intangible assets to help companies scale reaches its zenith when a business is drawing on artificial intelligence. AI is the ultimate intangible asset, because it takes on its own momentum – the algorithms give you more and more value without you having to do very much. The cycle looks like this: You feed data into an AI and it becomes more effective – tailoring a product to your needs, perhaps recommending news stories you want to read or songs you want to listen to. This improved service becomes more desirable, and so more of us use it. As more of us use it, we generate more data about our tastes and preferences. That data can then be fed into the AI, and the product improves.[26] This is another type of network effect – the 'data network effect', defined as 'when your product, generally powered by machine learning, becomes smarter as it gets more data from your users'.[27]

Take Google. You might ask why the search engine is so dominant. It wasn't the first to market: that accolade likely goes to WebCrawler, a search index founded in 1994.[28] Google was, however, the first search engine that effectively captured data network effects.

Google uses data about which links you click on (and which you don't) to improve the quality of its rankings. Every time a user clicks on a site and stays there, Google's AI learns that the link is useful; every time a user ignores a link, or clicks it then promptly hits 'Back', Google learns it is more of a dud. And so every click feeds into the algorithm.

Writing in 2008, Tim O'Reilly, one of the most astute observers of Google, argued that 'Google is the ultimate network effects machine'.[29] But data network effects are not unique to Google. They are common among the companies of the Exponential Age. When Netflix gives you a recommendation for a TV show, that recommendation is a result of your viewing habits compared to what other people are watching. Those choices are fed into Netflix's systems to improve the suggestions you receive.

Data network effects are a bit like a perpetual motion machine for intangible assets. The more these effects operate, the more they give. Companies that master this cycle can use intangible assets to take on an unassailable position. In 2020, Google remains completely dominant in online search – in spite of the attempts of major rivals, like Microsoft's Bing, to break into the market. It is a neat example of how in the intangible economy, firms can scale to a once-unimaginable degree. When a company is trading in intangibles, it need not worry about building new factories, or extending its supply chains. In some cases, it can just feed more and more data into its own system – and growth takes care of itself.

We can start to make out the distinctive characteristics of the exponential economy. This is a world in which network effects are key: once a company is gaining customers, it grows exponentially, leaving behind all of its rivals. This ruthless dynamic is driven by the emergence of a new wave of platforms, in which network effects are unusually powerful. And it is further catalysed by a wider economic shift towards intangible assets – which can scale more easily than their physical-world forebears.

The emergence of Exponential Age superstars has fundamentally rewired the logic of the global economy. As we saw at the beginning of this chapter, in the twentieth century companies would eventually reach a limit in their growth and would become large, mature 'blue chips'. When researchers looked at nearly 29,000 companies in 2010 they found that 'all large mature companies have stopped growing'.[30] But among superstar companies, this inherent tendency towards stagnation no longer seems to hold. The brakes that slowed the growth of companies have been snipped.

The most valuable firms in the world – Apple, Google, Microsoft, Facebook, Tencent, Alibaba and Amazon – are all Exponential Age companies. Only one of the world's largest firms, at time of writing, is a 'traditional' business – Saudi Arabia's state-run oil giant, Aramco. And despite their scale, the digital giants' revenues are still growing at a clip we'd associate with much smaller, younger firms. Take a single, reasonably representative year, 2018–2019. The Chinese tech giant Tencent grew its revenues by 18 per cent; Amazon by 20 per cent; Facebook by 27 per cent – all far quicker than the rate of growth of the global economy. The following year, the companies grew by 28 per cent, 38 per cent and 21 per cent respectively.

These companies have inverted one of the basic laws of business: diminishing returns to scale. They have learnt to push past the law of gravity that once limited a company's size. For superstar companies, the return on their investment grows as they get bigger. Ours is the first-ever age of increasing returns to scale.

Let's take a seminal example, from the US technology sector. Salesforce is a company that makes software for salespeople. It was founded by Marc Benioff, a charismatic software executive, during the wintery aftermath of the dot-com bubble. Benioff realised that the increasing power of the internet would let him deliver his sales-tracking application through a web browser. Rather than an upfront purchase, customers would rent the software.

It seems so obvious to us today, but in 1999 it was a radical idea. Back then, corporations bought software, installed it on central servers, and then had IT workers wander the offices to install the application on every single desktop computer. It was a pain. Benioff pioneered the idea of 'software as a service' powered through a web browser – and in the process he transformed the software industry.

In 20 years, Salesforce has grown from a handful of founders to a $200-billion business with nearly 50,000 employees. But as it has expanded, the company has figured out how to make each additional employee more productive – and how to get more money from its customers. In 2005, each of Salesforce's then 767 employees supported $230,000 in revenue. In 2020, each of its 49,000 employees supported nearly $350,000 in revenue. Customers seemed to love Salesforce too. In 2010, the average customer paid Salesforce $18,000. By 2020, that had risen sixfold. Not only were customers paying six times as much, they were happier. In 2010, 81 per cent of customers renewed their business with the firm. By 2020, more than 9 in 10 did.[31]

This is the reckoning of the forces identified above. The exponential logic of network effects means that whatever sales platform is ahead becomes completely dominant, provided executives don't take their eyes off the ball. The easily scalable nature of intangible assets means that Salesforce can grow and grow, without taking on that many more staff – or becoming unwieldy.

Netflix has followed a similar pattern. In 2010, Netflix had sales of $2.1 billion and 2,100 employees. Each one of them could account for slightly less than $1 million in revenue. A decade later, the firm had grown more than 10 times in size, with revenues approaching $25 billion. Its 9,400 employees counted, now, for nearly $2.7 million in revenue each. The dynamics of increasing returns is reflected in Netflix's content production, as well. In 2011, 14 years into its existence, Netflix had zero original content titles under its wing. In 2012, it launched one. By 2019, it released more original shows and films than the whole TV industry's yearly output prior to 2016.[32] With 37 Oscar nominations under its belt in 2021, the company is close to breaking a 1940 record of 40 nominations for a single studio. In these moments, it becomes apparent how increasing returns to scale aren't just rewiring business, but also culture and wider society.

Exponential Age companies are all too aware of the dynamics of this new economy. These firms have come of age in a peculiar environment: in each sector, one company dominates and everybody else gets left behind. As a result, these companies have developed a winner-takes-all mindset that reflects a winner-takes-all market.

Consider the management structure of many new superstar companies. The economist W. Brian Arthur was among the first to spot that a shift in organisation design was underway. In 1996, after studying the success of Microsoft and its operating system platforms, he realised that businesses based on network effects operated differently to traditional firms. Rather than chasing a few points of growth every year, they would chase large rewards. For these companies, he wrote, 'managing becomes redefined as a series of quests for the next technological winner . . . the Next Big Thing.'[33] At the same time, these firms' managers developed an obsession with growth. They knew that if they stopped growing, they could soon cascade into irrelevance – in the exponential economy, only one firm in every sector can triumph.

This emphasis on growth has forged a new vernacular within Silicon Valley. Reid Hoffman, who co-founded PayPal and LinkedIn, is fluent in this language. One afternoon in 2016, over mineral water in his offices in Sunnyvale, California, he gave me his new explanation for how tech start-ups grow quickly – 'blitzscaling'. Entrepreneurs who understand blitzscaling, he said, approach their companies very differently to the older titans of industry. Companies that blitzscale emphasise growth over efficiency. Practically, this means throwing out the traditional manager's rule book of optimising spending. Instead, aim for growth, even if it is expensive to do so. Peter Thiel, one of Hoffman's co-founders at PayPal, also has an ideology that focuses on growth. In Thiel's view, 'competition is for losers'.[34] He recommends instead that founders identify markets where they are substantially better than any existing competition – so they can gobble up market share in a way that would have been unthinkable just a few decades ago.

What Thiel and Hoffman identify, in their respective ways, is that the superstar firms of the Exponential Age don't just aspire to become large and dominant. They have no choice: second place is a very, very distant second.

In practice, these firms' obsession with constant growth manifests in three different ways. First, there is 'horizontal expansion'. This is when firms move from one market to an adjacent market. For example, the story of Alibaba, a Chinese marketplace for retailers, is one of never-ending expansion into new areas. It has operated an online payments business called Alipay since 2004. By 2013, Alipay had become the world's largest mobile payment service, and its parent company spun the business out as Ant Financial. This was the first step in horizontal expansion: giving this new finance company its own autonomy. Ant Financial drew upon the huge amounts of transaction data Alibaba had been collecting, channelling the power of data science – through what it described in hyperbolic marketing speak as the 'Ant Brain' – to become an ever more dominant force. It was a classic example of the power of network effects: the more data Alibaba had, the more powerful and effective it became; the more powerful and effective it became, the more customers joined up – and the more data it had.

The logic of network effects and increasing returns led Ant Financial to expand into increasingly distant portions of the market. For example, by looking at purchase data the Ant Brain revealed that women who bought skinny-fit jeans were more likely to spend money on smartphone repairs. The company's executives hypothesised that phones were slipping out of tight back pockets and landing on the ground with a smash. They quickly launched a screen insurance product that was aimed at young women who bought fashionable trousers.[35]

The result of this kind of expansion was Ant Financial's customers using myriad different parts of the company's service, across their economic lives. Ant Financial soon expanded to include five businesses: payments, wealth management, credit scoring, lending and insurance. It has persuaded 40 per cent of its customers to use all five products, and 80 per cent of its customers use three or more.

This horizontal expansion is hardly unique to China. Apple, which started making computers, now makes phones, tablets and watches. It has a media subscription business and an App Store. Its products provide health information. And Amazon, a retailer, has moved from selling books to peddling virtually every type of tchotchke. It has a logistics wing, the world's largest cloud computing business, and – like Apple – a media business.

But nor does exponential logic lead solely to horizontal expansion into new markets. The second form of superstar growth is vertical expansion. This is where a firm looks at the activities of part of their own supply chain and decides to bring them in-house. Vertical integration is not unique to the Exponential Age, of course. Andrew Carnegie became the second-wealthiest man in the United States in the 1800s when he expanded his steel company so it took over every stage of the supply chain – taking control of iron mines, coal mines, railroads and steel mills, all to achieve enormous efficiency in the steel industry. However, by the late twentieth century, vertical expansion had gone out of fashion. Instead, management orthodoxy was that companies should stick to their 'core competencies'.[36]

But not any more. As we marched into the Exponential Age, the spirit of vertical integration returned. Google's advertising business, responsible for virtually all the firm's revenue and profits, is a vertically integrated system. Several different mechanisms make online advertising possible: targeting, auctioning the ad space, placing the ad, tracking the engagement. Once upon a time, different companies looked after each of these processes. By 2020, an advertiser wanting to buy an ad to show to a particular group of users can do so entirely with Google's products. Google's acquisition of YouTube in 2006 helped, vastly increasing its dominance in the advertising ecosystem; so too did the launch of Maps, which helped advertisers understand even more about where their consumers were going about their business. And this vertical integration shows no signs of slowing down. In 2010, Google acquired 40 companies – a diverse portfolio of firms that specialised in social gaming, photo editing, computer processors, touch typing, voice recognition, travel and music streaming. All of these acquisitions allowed Google ever-greater control of its own supply chain.

Such vertical integration can even involve expanding from the massless, digital world into creating physical hardware. Take chip making. For most of the history of the personal computer, chips were made by one specialist firm, the computer by another and its operating system by yet another. Today, however, the big computer companies are making their own chips. Starting in 2010, Apple began to make chips for its mobile phones, and in 2019 chips for its computers. In 2018, Google started to make its own specialist chips to speed up AI operations. Even Tesla, the car company, ditched commercially available chips in favour of its own creations in 2016.

Finally, in addition to expanding into pre-existing markets, these superstar companies have the budgets to invent new sectors of the economy from scratch. One of the boons of increasing returns to scale is that exponential companies can invest heavily in research and development – as we saw with Amazon's staggering research budget described in Chapter 3. In fact, the biggest Exponential Age firms increase their R&D spending faster than their revenues. The goal here is to make technological breakthroughs that allow these companies to foster completely new sectors – which they can then go on to dominate.

In the case of Alphabet, which owns Google, the vast bulk of revenues come from search engine advertising. Yet Alphabet's researchers are active in domains light-years from search engines. The company has a team investigating 'cold fusion': the prospect of harnessing the nuclear fusion processes that occur within stars, except at room temperatures.[37] If they cracked cold fusion, Alphabet would be able to transform the way the world accesses energy. And it's not just energy. Google's R&D spin-off, X, has funded experiments in manifold sectors – the life sciences, consumer internet access, self-driving cars, cyber security, machine learning. The group even launched a new business called Malta, which stores energy cleanly by heating huge vats of salt past melting point.

All this confirms that the superstar company is not a fleeting trend but a wholly new economic paradigm. The corporate giants of the future will take increasingly dominant market positions, both within and across sectors. But is this necessarily a problem? A glance at recent history would suggest the answer is 'yes'. In the twentieth century, dominant market positions tended to lead to monopolies. And monopolies were an issue. In the absence of competitors, companies might be tempted to offer customers a worse deal. Traditional monopolists often drove prices higher and took a laid-back approach to improving products and services – they might even use devious methods to prevent competitors from entering a market.

In the Exponential Age, troublesome dominant firms are more likely than ever – but the issues they pose are subtly different. They require a rethink of the way we think about monopolistic practices. In the 1970s, Robert Bork – the solicitor general of the United States under presidents Richard Nixon and Gerald Ford – came up with what became the standard approach to antitrust law, the body of US legislation concerned with monopoly. Bork's emphasis was less on anti-competitive practices or cartel-like behaviour per se, and instead on consumer welfare. The question, in Bork's view, was: will a company's behaviour hurt a consumer's pocket? A company's size or market position was not, in itself, an issue – the implications for consumers were what mattered.

With this mindset, Bork – whose ideas were influential in both the US and Europe – may not have seen Exponential Age superstars as a problem. In the technology and internet sectors, it is far from clear that customers are getting a bad deal. Sure, Google and Amazon are tending towards monopoly in some sectors – but consumers aren't being ripped off, at least not at first glance. The gift of exponential improvement means that the technology products improve every year for a fixed cost. That $1,000 computer you just bought is better than the one you bought for $1,200 three years ago. An internet search is free – and today, searching the internet is faster and more effective than it ever has been. Cameras in phones and easy-to-use sharing applications have brought the joy of photography to billions of people for practically nothing. In every area of the digital economy, the consumer experience seems to be getting cheaper and more efficient. So much for the perils of monopoly.

But there's a catch. Robert Bork's framework for understanding monopoly doesn't account for the real issues created by monopolistic companies – at least not in the Exponential Age. It is an example of the exponential gap. There are problems with the new age of monopoly, but these problems don't manifest themselves in the way our existing norms and rules can grasp.

The first problem is that, instead of exploiting consumers, modern monopolistic businesses might exploit smaller-scale producers. Consider the App Store, through which many of us buy interesting apps for our iPhones. Apple, which makes the iPhone, also operates the App Store. And the company charges software developers to sell their products there. For small software developers making less than $1 million a year, those fees are 15 per cent per sale. For larger ones, 30 per cent. As of 2020, the platform hosts over 1.8 million apps, with gross sales in excess of $50 billion a year.[38] But is that 15 (or 30) per cent take fairly earned? Or is it just the hefty protection money one has to pony up to the neighbourhood enforcer?

Apple clearly does something for its developers. It provides them with access to billions of people who use iPhones; it provides a secure and trusted environment for people to buy apps, which encourages us to buy; and it makes the buying process easy, handling payments and even subscriptions. Before the App Store, the mobile app market was moribund.

Finding apps was hard, the payment process was clunky and buggy, and insecure software was commonplace. Apple's creation of this marketplace is what created the market.

But are Apple's efforts worth 30 per cent of a developer's efforts? Many developers don't think so. Epic Games, which makes Fortnite – a video game beloved of teenagers and loathed by their parents – was one.[39] After challenging Apple's fees and attempting to bypass the 30 per cent cut, they were kicked off the platform. (Apple did eventually back down, sort of – a few weeks after the fracas, they halved fees for some developers, albeit only for those with much lower revenues than Epic.) The problem is that we can't really know what a fair price is in the absence of a competitive market. Given that Apple has an 80 per cent market share among the young people Epic targets, you can't just turn to the market to work out the 'true' value. In some segments of the economy, Apple is the market.

Epic is a huge company, and perhaps we need not shed too many tears for it; there are, after all, games other than Fortnite that people can play. But this phenomenon – of oligopolistic practices that hurt sellers rather than consumers – is common across platforms, and not all of the victims are so mighty. Consider Google's vertical integration of its ad business. Google's dominance of the online advertising industry has led to questionable business practices. The company has been accused of setting price floors that lead advertisers to pay more than they need to, as well as withholding information of value from the advertisers. The European Commission fined Google €1.49 billion (approximately $1.8 billion) for abusing its dominant position in the ads market, finding among other things that the firm did not allow its competitors to sell ads in Google search results.[40] The UK's Competition and Markets Authority also voiced concerns over Google's vertical integration in the advertising market, noting that the business was rife with conflicts of interest – and the strong risks of fee increases, given the absence of effective competition.[41] Again, however, it's not the average Google user who loses out – it's the small business that needs to buy ads.

This leads to the second problem in this new age of monopoly. Economies that are dominated by large companies become progressively less dynamic. Bork's theory of monopoly also struggles here. His framework accounts for price hikes within a specific sector, but not a wholesale loss of dynamism across an economy. But there are signs that just such a loss of agility is taking place in the Exponential Age.

At first glance, this idea seems absurd. After all, many Exponential Age winners are young companies, often less than 20 years old. If markets were getting less competitive, how could these firms have grown so quickly? And if markets are dominated by digital leviathans, why are more start-ups being created and being more well-funded than ever before? Globally, more than $288 billion of venture capital was invested in new start-ups in 2020 – a 15.4 per cent annual growth since 2001's level of $19 billion.[42] And start-up creation is becoming less and less concentrated in the US. In 1995, virtually all venture capital flowed to American start-ups, but by 2020 about half of all investment went to companies outside of the United States. Venture capitalists are not fools: if markets were as locked down and incontestable as the winner-takes-all framing might suggest, why would they back the creation of new companies? Ours looks like a dynamic, vibrant economy – superstars be damned.

But we need to be more careful. For even as new companies spring up, very large firms are increasingly adept at cementing their own position in that market – by buying nascent competitors when they are young. Google bought YouTube in video, DoubleClick in advertising and Android in mobile phones. Facebook picked up WhatsApp in messaging, Instagram in social media and Oculus in virtual reality.

This has long-term consequences for innovation. Research shows that breakthrough inventions are more likely to come from individual inventors or smaller teams – one group of Ph.D. researchers analysed 65 million papers, patents and software products from 1954 to 2014, and found that 'while large teams do indeed advance and develop science, small teams are critical for disrupting it'.[43] And while, as we've seen, a few companies take a very ambitious approach to research – Google springs to mind – the data suggests that those corporate giants can also come to narrow the focus of research in fields where they dominate. Researchers who analysed 110,000 papers in the broad field of AI concluded that much of the research conducted by companies tended to be very narrow compared to academic research. Commercial research, which seemed to be laser-focused on approaches which had already been shown to be successful, tended to reduce the number of pathways being explored.[44]

At their most audacious, tech companies even end up diverting scientists' attention away from big-picture scientific research. In the 2010s, AI skills were in such demand that the tech giants started to poach university professors to help with their corporate priorities. The goal was both to access the universities' expertise and to create a pipeline of recent graduates to follow their teachers from academia into corporate AI labs. Between 2004 and 2018, North American universities saw 221 AI professors leave academia – the majority to pursue a career in the computer industry or start their own business. During the same period, 160 AI faculty members moved to another university.[45] These numbers may seem low, but as the leading AI researcher Stuart Russell told me: 'We don't have a lot of AI professors and they take a long time to make.'[46] Between 1997 and 2006, AI academics were 100 times more likely than life scientists to move from academia to industry.[47] With the brain drain from universities to the private sector, research is at risk of becoming more and more narrow – increasingly focused on the commercial priorities of big companies.[48]

And there's a final problem with the tendency towards monopoly. Bork doesn't mention it much, but companies are not just important to consumers – they are also part of the wider functioning of society. Most obviously, they are supposed to pay tax. And for the first giants of the Exponential Age, our tax codes have been generous. The bulk of the assets for Exponential Age firms reside in intangibles, and intangibles are prone to sliding across borders – and easier to slip past the taxman.

According to *The Economist*, in the years to 2020, the largest American tech firms paid a tax rate of around 16 per cent of their profits, far below the then 35 per cent corporate US tax rate.[49]

Their methods can be ingenious. One loophole, long beloved by the American technology titans, was the 'Double Irish', in which large companies put their intangible intellectual property in an Irish-registered company controlled from a tax haven such as Bermuda. Ireland considers the firm to be offshore, but the US considers it Irish.[50] And the profits? Well, the profits remain untaxed. This practice has started to go out of vogue, and after the Organisation for Economic Co-operation and Development (OECD) and Donald Trump's administration clamped down on the Double Irish, at least one technology company brought it to an end in 2020.[51]

In general, governments have been caught off guard by the rapidity with which the Exponential Age titans have grown. A few nations, like Ireland, spotted the shift. The country was quick to use appealing tax rates as a way of enticing the new economy superstars to set up shop there.[52] But many countries did nothing, standing by as their leading companies relocated and starting paying less tax.

Yet for all their difficulty, the problems posed by the ascendance of Exponential Age superstars are not intractable. In fact, we have the ability to answer them already. It just needs a change of mindset. Above all, we need to develop a new way of thinking about monopoly. In the Exponential Age, Bork's theory is in tatters: we won't necessarily be able to spot over-large companies from their negative effects on consumers. But fortunately, the market dominance of big tech companies has given the once-fusty field of antitrust scholarship a new lease of life – and has led to some exciting new ideas.

For example, one surprisingly popular paper by the legal scholar Lina Khan argues for an alternative antitrust framework – one that accounts for the implications of Amazon's dominance of infrastructure and logistics.[53] Instead of focusing on Amazon's implications for consumers, we might look to the conflicts of interest that arise from massive scale and infrastructure power. Consider the case of Ecobee, which makes gadgets that control household heating and sells them via Amazon's website. Consumers can connect them to Amazon's Alexa voice assistant. The *Wall Street Journal* reported that Amazon threatened to hamper Ecobee's ability to sell, unless it shared consumer data with the behemoth. According to Khan, this kind of behaviour – rather than size per se – would be a key indicator of monopoly. Khan's suggestions are nascent, but they are apposite. Once we have a new set of rules which deal with large companies in place, it will become somewhat easier to prevent the superstars wielding too much power.

But what should we do when our biggest companies do start to creep towards monopoly? For one thing, antitrust bodies need to be more confident about blocking big companies from acquiring smaller ones. The trouble for any regulator is determining when an acquisition by a dominant firm could result in monopoly in five or ten years' time. Google paid peanuts for the mobile system Android in 2005. It was a time when the phone market was dominated by a battle between BlackBerry and Nokia – what harm could Google buying a start-up do? Facebook bought Instagram when it had a dozen staff, for a sum that many at the time thought was absurdly overinflated. In neither case did the US's antitrust authorities realise that, in a decade's time, both services would be gargantuan.

The solution, perhaps, is for regulators to reserve the right to approve any acquisition by the largest companies, even of relatively small players. They might also be empowered to insist that an acquisition be sold off at some future point if it turns out to be troublesome.

This approach would stop Exponential Age platforms buying up areas of the market. But we might also seek to stop these companies growing so large organically. Here, a valuable idea is 'interoperability'. We encountered this concept earlier, in the shared language that allows different types of email account to interact with each other, which helped drive the early growth of the internet. Today, interoperability between platforms is all too rare. If you are a driver on Uber and you decide to move to a rival service, you will start there from scratch – without any customer ratings or reviews. It would be a little like getting a new fax machine and having to persuade all your clients to buy a machine of the same model so you can interact.

Insisting on interoperability, especially for firms that get to a certain size, is one way of tampering the network effects that enable companies to get ever larger. For example, if I listed my goods on eBay, someone could find them on a rival service and bid for them. If I posted a short video of my weights routine on Instagram, my friends on LinkedIn should be able to admire it and send me digital plaudits. Interoperability preserves the positive benefits of the network effect: even if I leave one particular social network or other service, I can still access users on the previous service. But it also means that the network effect on any one service – leading a single company inexorably towards monopoly – is curtailed.

We have been on the journey towards interoperable communications services before. In the early 1990s, mobile phone companies were in silos. You could only send text messages to people who were also subscribers to the same service. If I used Vodafone in the UK but a friend used the Orange mobile network, I couldn't send her a message. There was no real technical reason for this incompatibility, merely a business one – each network worried about its customers going over to its rivals. This meant that in 1999, before this type of compatibility was widespread, the average British mobile phone user sent a couple of text messages a month. By 2004, four years after interoperability became commonplace in the industry, the average Brit was firing off 33 messages a month. By 2010, that number was 141.[54] The ability to freely access subscribers on any network, coupled with declining prices, drove up phone usage. And the fear of interoperability proved misplaced – the companies were stronger than ever, and customers' experience was vastly improved.

Today, we are taking steps towards making interoperability a reality across sectors. Since 2012, the Financial Conduct Authority, Britain's financial regulator, has insisted that consumer banks make it easier for customers to switch banks. More recently, the movement towards 'open banking' forces banks to share their customers' data (with a customer's permission, of course) with other apps. This is why a Brit can often move from one consumer bank to another with relatively little hassle – compared to 20 years ago, when you'd have to show up to a stuffy office in person with photocopies of documents that you'd already provided to someone else. The European Commission is taking the lead in this area: its Digital Markets Act, proposed in December 2020, puts new obligations on 'gatekeeper platforms' – the type of dominant network platforms we've been discussing – including compulsory interoperability between platforms.[55]

And finally, we can limit the power of gargantuan Exponential Age firms by treating them less like ordinary companies and more like utilities – the essential services that we can't avoid, like water, or the electricity network, or sewage pipes. Increasingly, digital firms fit this mould – they have become an inescapable part of our lives. Try going a day without accessing the services of Google, Apple, Amazon, Facebook, Microsoft or Netflix (or your local equivalents – Tencent or Alibaba's services in China; Yandex in Russia). You will discover this is near impossible. It is hard to participate in the local economy, let alone the global one, without a smartphone – and without access to an app store. Neither domestic nor school nor professional work is really possible without access to search engines. Participation in society is easier if one uses social networks. But with this power, whether the companies like it or not, comes a responsibility.

We have long placed stringent demands on utility providers. Regulators usually demand higher standards from them. For telecom companies, this meant, for example, the 'universal service obligation' – meaning every resident of a country must be offered reasonable telephone services. Others are treated as essential facilities and must share their infrastructure with competitors in a fair way. The UK's phone company, BT, is a good example. Its phone network is a natural monopoly which made it hard for rivals to compete – they would be dependent on BT, their competitor, for access to the wire that leads into customers' homes. After many years of tussling, the British regulator came up with a solution: BT's backbone would be split off into a separate company, still owned by BT but operating at arm's length. This firm, Openreach, would need to invest in the network and sell services to BT and any other phone company who wanted access on equal terms. This eliminated the need to build a second national network into people's homes (although it didn't stop some firms from doing so).[56]

Again, the European Commission is leading the way in regulating big tech companies. Under the proposed Digital Services Act, when a platform gets very big – defined as reaching more than 10 per cent of Europeans – new obligations come into play. These involve greater reporting and auditing requirements, and a commitment to share data with researchers and the authorities. These more onerous requirements do not apply to smaller companies – meaning that, hopefully, they don't calcify the wider market.[57] But if we are to truly treat digital platforms like utilities, and attempt to break their chokehold, we may need interventions of this kind.

All of these policies are about closing the exponential gap. It is almost a gospel truth that markets tend to remain fairly competitive. Yet in the Exponential Age, that is a dangerous assumption. Politicians, businesspeople and regulators haven't yet realised that we are entering the age of the winner-takes-all market. They imagine that, if monopoly does rear its head, it will lead to negative outcomes for consumers – but this is no longer the case.

It is an object lesson in the slowness of institutional change. Superstar companies have been on the horizon for a long time – as we have seen, W. Brian Arthur wrote about the issues of platform dominance and increasing returns to scale in the tech world in 1996. When Arthur first wrote on the subject, however, it represented an interesting observation about a niche industry. Today, this has become the most important and largest industry in the world. In their attempts to regulate it, governments ended up applying industrial-age thinking to Exponential Age problems.

But this twentieth-century world view can be overcome. Governments will, by their nature, operate more slowly than the fastest-growing companies. Yet between the worlds of academia and jurisprudence, think tanks and business forums, the policies that will close the exponential gap are at our fingertips. In every sector, people increasingly recognise that the era for which we made the rules has ended. And so the rules themselves must change.

5

Labour's Loves Lost

At some point in the middle of the 2010s, the media suddenly got very concerned about the effects of automation in the workplace. 'Millions of UK workers at risk of being replaced by robots', blared *The Guardian*.[1] 'Will robots take our children's jobs?' asked the *New York Times* in 2017.[2] In 2015, an apocalyptically titled book by the futurist Martin Ford, *The Rise of the Robots*, became a bestseller. Its hypothesis: automation has the potential to create a 'jobless future' for future generations.

Such fears have a long history. Since the time of the industrial revolution, employers have used technology as a way of squeezing an extra bit of productivity from their factories and their workers. And this has often created a tension between employers and staff. Machinery can look attractive to a company owner: they can replace a human, who would need a regular wage, with a one-off payment and a few running costs for a machine. As the eighteenth century gave way to the nineteenth, workers' anger at machines taking their jobs grew. In 1810s England, the Luddites – a rag-tag group of disenfranchised workers who took their name from Ned Ludd, a mythic figure who supposedly declared war on his employer's machines – began destroying the automated looms that were supposedly ruining their livelihoods. Over the next couple of centuries, similar movements would crop up around the world: 160 years after the Luddites, General Motors workers in Lordstown, Ohio, set fire to assembly-line controls, forcing production to halt. They were protesting against increasing automation, which they claimed was resulting in layoffs and an inhumane increase in what managers expected of workers.[3]

In other words, we've seen the 'rise of the robots' before. Anxieties like these usually crop up during periods of rapid technological change. And it's not just the workers themselves. The economist John Maynard Keynes popularised the idea of 'technological unemployment' in 1928, when he noted, 'The increase of technical efficiency has been taking place faster than we can deal with the problem of labour absorption' – that is, the ability to find work for workers.[4]

Now we have entered the Exponential Age, new technology appears to be challenging workers' *raison d'être* once again. Artificial intelligence in particular seems to encroach on employees' working lives more and more. And, as we discovered in Chapter 1, these systems are rapidly becoming more advanced. If you want to recognise a person from a photograph, you are better using an AI than an employee. If you want to translate quickly between two languages, an AI will do it better than you. If you want to schedule shifts for 50 different workers, an optimisation algorithm will thrash any person's timetabling attempt. And with every year, these algorithms get better and better at tasks we used to think of as quintessentially human.

A more open question, however, is what this actually means for workers. The threat of automation looms large in our collective imagination. The rise of newly automated workplaces raises the prospect of mass redundancy. And it is framed as a more existential threat than Keynes's fears of technological unemployment. Soon, we are told, we'll reach a point where automated systems will render most of us unemployed and unemployable. In 2016, for example, Geoffrey Hinton – one of the AI pioneers we met earlier – publicly mused on the prospects of radiologists, the specialist doctors who deal with X-rays, computerised tomography and magnetic resonance imaging scans. Radiologists, Hinton told a small crowd of AI researchers and founders, were 'like the coyote that's already over the edge of the cliff, but hasn't yet looked down, so doesn't know there's no ground underneath him. People should stop training radiologists now. It's just completely obvious within five years, deep learning is going to do better than radiologists . . . It might be ten years, but we got plenty of radiologists already.'[5]

Was Hinton right? On one level, yes. In the five years following his comments, artificial intelligence systems – specifically those relating to machine vision, the subfield most relevant to radiology – improved as he had predicted. A childhood friend of mine, Rajesh Jena, is a consultant neuro-oncologist at the University of Cambridge Cancer Centre. Together with researchers at Microsoft he has developed a tool that provides accurate 3D visualisations of tumours and organs. This has cut down the time it takes to identify a neuroblastoma from a few hours to four minutes, allowing the specialist more time for calming explanations to nervous patients.[6]

But at the same time, Hinton was wildly off the mark. Consider what actually happened to radiologists' numbers. In 2010, before the present deep learning wave, there were 27,986 radiologists in the US. In 2015, the year before Hinton's talk, there were 27,522 radiologists. But by 2019, three years later, the number of radiologists had risen to 28,025.

One could argue that three years is too short a time frame to track the impact of this technology. But Hinton had been explicit: 'People should stop training radiologists.' That would have been unwise. Far from being out of work, radiologists remain in high demand and in short supply. And that is in the highly developed health system of the United States. Most of the rest of the world faces a shortage of both radiologists and the machines that they need. In practice, the kinds of tools that Rajesh has created seem to be helping overworked radiologists rather than making them redundant.

This example hints at a more complex, nuanced vision of the future of work. It is true that automated systems do increasingly excel at once-human tasks. Store-cleaning robots are slowly becoming more common in the US, especially since the start of the Covid-19 pandemic.[7] In China, machine vision systems scan photos of damaged cars to evaluate the likely cost of repairs; no human needs to look at the vehicle.[8] The list goes on.

But the notion of a jobless future – the 'robopocalypse' of tabloid headlines – is overstated. It is alluring and it makes the news, yet it is muddle-headed. Historically, our economies have become more automated. And historically, employment levels have tended to increase.

How is this possible? Because automation has the potential to create more work than it destroys. Yes, there may be unemployment in the short term as some sectors wane. But automation creates jobs which often require new and distinctly human skills: from the programmers who develop systems to those who operate and maintain them. Over time, automation will end up inventing whole new sectors of the economy – ones that we can now only imagine.

The snag with the 'rise of the robots' narrative is not that it is merely wrong, however. It is also a distraction. We are living through one of the greatest transitions in the history of work. Technology will revolutionise how we all work, and the relationships between employers and employees. And this transition, if handled badly, has the potential to cause a new era of workplace exploitation.

But the problem is not automation per se. Our economies will continue to create new work to be done, and so the quantity of jobs will likely remain high. But the quality of jobs – the work an employee has to do, the regularity of their income, their opportunities to build skills, the say they have in their working conditions – could precipitously decline.

We are witnessing the emergence of an exponential gap between the working arrangements enabled by rapidly improving technologies and the outdated norms, regulations and expectations that govern working life. As we'll see in this chapter, the shift to platform business models creates new ways for companies to organise work: with ever-greater emphasis on gig-working rather than traditional employment contracts. The use of sensors and analytics allows employers to implement automated management systems designed to boost efficiency and productivity – often at the expense of workers' wellbeing. And all the while, the share of value going to the workforce – rather than company owners – is declining.

None of these three forces can be captured by an idea as simple as the robopocalypse. But each is driving a wedge between those who can harness the power of the Exponential Age and those who will be left behind.

The notion of mass automation and job losses hangs over Silicon Valley like the low-lying San Francisco fog. While the argument is an old one, it re-emerged in the early years of the twenty-first century – as the advances in AI that we explored in Chapter 1 started to come to fruition. Bookshops were soon awash with books called things like *The Second Machine Age* and *Surviving AI* – all making, in one way or another, a devilishly simple argument: artificial intelligence changes the game. We have a generalised technology that is adaptable; it can be used in many different circumstances. And it is adaptive; it learns from experience. As a result, it will continually improve over time, making large chunks of the workforce redundant.

After all, AI breakthroughs were occurring in some of the most human-seeming of skills: perception, communication, planning and manipulation. What's more, researchers were showing off machines that could perform these hitherto unautomatable tasks with greater aplomb than the most skilled human. Surely that would lead to job losses, with software and robots replacing people.

In the most famous piece of research on the topic, two Oxford academics, Michael Osborne and Carl Frey, predicted that as much as 47 per cent of the US workforce were in jobs at risk of redundancy thanks to advanced computerised systems such as machine learning.[9] Forecasters and futurists leapt onto these and similar findings – Osborne and Frey's research was cited more than 7,000 times in seven years.[10] In one case, in 2017, the market research company Forrester predicted that nearly 25 million US workers would lose their jobs due to automation by 2027, and that automation would only create 14 million new ones.[11] A BBC headline warned of 20 million job losses globally by 2030.[12]

This narrative holds more than a nugget of truth. In the 2010s, many people did lose their jobs to automation. In 2017, the boss of Deutsche Bank talked of using automation to get rid of thousands of jobs, especially people who 'spend a lot of the time basically being an abacus'.[13] And though banking is a particularly unsentimental industry, Deutsche Bank weren't alone in wanting to automate office tasks. Company after company launched initiatives to eliminate white-collar labour. The demand was so great that it created a boom in office automation. One start-up that made automation software was a Romanian firm called UiPath. In 2015, it was still a tiny business, with fewer than 20 employees. Over the next five years, large firms around the world turned to UiPath to automate white-collar work, and the company mushroomed to more than 3,000 employees. By 2020, it was worth more than $35 billion.[14]

With automation gathering pace, the argument that its effects would show up in the employment statistics seemed indisputable. If an employer can increase their productivity using new technology, they'll end up employing fewer people – won't they? Long-term data seemed to back this up. Employment in the American manufacturing sector peaked in absolute terms in mid-1979, when 19.5 million Americans held jobs in the sector.[15] American workers were the most productive in the world, half again as productive as the British worker and a fifth more productive than the German worker.[16] Over the next few decades, manufacturing output continued to rise but factory employment declined. American factories were producing more stuff (as measured by the value) but needed far fewer workers to do so. According to the Brookings Institution: 'In 1980 it took 25 jobs to generate $1 million in manufacturing output in the US.' By 2016, it would take just five workers to produce the same value of finished goods.[17]

And this trend seems, at first, even more pronounced among today's tech firms. Consider the staffing levels at what were once some of America's largest firms. When it had its largest workforce, in 1980, General Motors employed 900,000 people. It sold more than 4 million cars a year, with revenues approaching $66.3 billion.[18] Each employee accounted for about $74,000 in sales. Now consider the biggest firms of the Exponential

Age. Alphabet, which owns Google, employed about 120,000 people in 2019, against revenues of $162 billion, clocking in at $1.4 million in sales per employee.[19] The overall trend is towards more valuable companies with fewer employees.

And yet there's a catch. Even as the tech giants grew, employment figures continued to look rosy. Until the Covid-19 pandemic of 2020 froze the world's economies, many countries were enjoying record levels of employment. Across the OECD, a club of 37 (mostly) rich nations, employment levels were at record highs as late as 2019 – higher than they had been since even before the global financial crisis of 2007–2009.[20] Globally, the picture was similar. The International Labour Organization estimated that, heading into 2020, global unemployment was the lowest it had been since 2009.[21]

The robopocalypse, it seemed, had been delayed.

We're faced with a conundrum. On the one hand, automation seems to be threatening a large fraction of the workforce. On the other, the better the technologies get, the more jobs there seem to be. What is going on?

There are a few possible explanations. The first is that automation is perhaps much harder than it seems – for all the talk of an imminent robot world, we're actually at a fairly early stage in the automation process. Although technologies are improving, many are taking longer than expected to become superhuman. Exponential processes take time to marinate.

This slowness is bound up with the fact that much work is harder to automate than you might think. The difficult-to-automate nature of many jobs is captured in a maxim known as 'Moravec's paradox' – first outlined by Hans Moravec, a professor renowned for his work on robotics and AI at Carnegie Mellon University in the 1980s. As he wrote in 1988: 'It is comparatively easy to make computers exhibit adult level performance on intelligence tests or playing checkers, and difficult or impossible to give them the skills of a one-year-old when it comes to perception and mobility.'[22] Over three decades later, Moravec's paradox still holds. We can build computers that can play Go, a game which has more combinations of moves than there are atoms in the universe. But there are a number of more human skills that computers can't cope with.

To see how this works in practice, it's worth looking at two sectors of the economy: one usually characterised, crudely, as 'high-skill'; the other as 'low-skill'. First, think of a Wall Street trader. The image – now a little outdated – is of a besuited man screaming across a trading floor, surrounded by screens stacked one atop another. His goal – and it is usually a 'he' – is to buy and sell shares or other financial instruments on behalf of clients. If you've seen *Trading Places*, you might remember the depictions of the frantic, animalistic exchanges in the trading pits of a commodities exchange.

But this job – coveted by generations of finance graduates – is easily automatable. Today, visit a trading desk and you'll see the pit-traders have largely gone, replaced by computers. When you buy or sell some shares, the likelihood is that those shares are sold via algorithm, finding the best price in the market. No humans needed. When I ran an innovation group at Reuters, the financial information giant, in 2006, algorithmic trading was starting to take off. About 30 per cent of all shares were traded that way. A decade later, nearly 70 per cent of shares were traded automatically.[23] The face-to-face trading of major financial instruments is increasingly rare.

Even fund management, superficially a more 'human' area of the financial services industry, is not immune to automation. For decades, fund managers took other people's money and personally picked investments they thought would get a good return. And for decades, this was a destination for the well-heeled international graduate who had a knack for both economic analysis and personal networking. These days, not so much. Fund managers are being replaced by automated systems. This move from human-run 'active' to automated 'passive' funds passed a milestone in late 2019, when more than half of all global assets under management went to simple rules-based funds rather than those run by people making investment decisions.[24]

This is all possible because the culture of trading – the screaming, the braces, the high-octane lifestyle – was mere theatre. A stock was a stock. A bid, a bid. An offer, an offer. Buying and selling shares was a matter of matching bids and offers. Managing all those assets – the patchwork of shares that make a portfolio – turns out largely to be done better by computer programs than by humans.[25]

But most jobs are not like Wall Street trading. They are more complicated than following a stock index or buying or selling the odd share. They are full of tasks that we don't bother to write down. Humans know how to open a door or socialise with the other humans in a workplace – and these human relationships are in large part what allows companies to function. Many parts of workplace interaction are governed by undisclosed codes that emerge as you interact with your co-workers.

The philosopher Michael Polanyi said that 'we can know more than we can tell'.[26] It is a delightfully human assertion that, all around us, there is knowledge that we can't put into words. This is the stuff we learn by just being around our co-workers, our boss, our customers. When we're embedded in an environment, we get cues and clues about how things get done – what really matters, who matters, what the trade-offs are, what the best shortcut is. Rarely is this written down – and even if it was, we would still likely learn this better from experience than by studying. There is a tacit dimension to our lives that is not codified, and perhaps never can be.

This is perhaps even truer of supposedly 'low-skill' jobs than it is of 'high-skill' ones like a Wall Street trader. The anthropologist David Graeber was fond of pointing out that many jobs that we sometimes deem repetitive, task-oriented and perhaps easily automatable are, in fact, more like care work. They're based less on specific tasks and more on human interaction and emotional labour. Think of a London Underground worker. In practice, this job isn't so much about watching ticket barriers as it is about helping other humans – guiding confused tourists, ensuring lost children find their parents, explaining to angry commuters why their trains have been delayed. As Graeber put it, 'It has more in common with a nurse's work than a bricklayer's.'[27]

This dynamic means that, in practice, much of the labour economists consider 'unskilled' might prove difficult to automate. A workplace manual is usually barely even a rough guide to what a job is – it doesn't cover half of the things you need to be successful. And when such tacit knowledge exists in a workplace, it makes artificial intelligence that can do the job very hard to build. An AI system needs a clear and unambiguous goal, and modern systems need to be trained on that data. If the know-how about a job is largely hidden, an AI system will be trained on only half the picture. In short: if an environment is designed for a person, it is likely to be too complicated for machines now or any time in the near future.

As a result, when automation does occur, it happens slowly and step by step. 'Work' has to be divided into smaller, more manageable pieces. A simple piece, perhaps the simplest, is broken off. A basic robot or piece of software can then tackle that elementary chunk of work. 'Simplification is mostly how automation happens,' the economist Carl Frey, co-author of the doom-and-gloom Oxford study cited above, puts it. 'Even state-of-the-art robotics would not be able to replicate the motions and procedures carried out by medieval craftsmen. Production became automatable only because previously unstructured tasks were subdivided and simplified in the factory setting.'[28]

That holds true even as artificial intelligence rapidly picks up speed. By the end of 2020, AI systems, in software or in robots, were not making a dent in the employment statistics. Automation is still only viable for relatively simple, usually simplified, everyday tasks. For example, the first AI-powered self-driving vehicles drove in the very measured environments of Phoenix, Arizona – with its wide, straight roads and perfect weather. The frantic, rainy autobahns of Germany will remain a tough ask for a while. When self-driving trucks are first launched, they will stick to straight highways, rather than manoeuvring through narrow roads in the City of London. Full automation, it seems, remains some way off.

Yet it would be a mistake to conclude that automation is a fringe force in the Exponential Age. As those Wall Street traders found, it is happening. And while, at the moment, its effects remain limited to specific tasks and sectors, we have no guarantee that it won't soon affect a wider array of jobs. As we saw in Chapter 3, making precise predictions in the Exponential Age is a fool's game.

But if automation is going to disrupt our work, it seems unlikely to be in quite the way that the headlines imply. Far from causing mass unemployment across sectors, automation instead seems to be a tool that companies can use to become more competitive. Those who use automation well, compete better – at the cost of less competent firms.

The Covid-19 pandemic provided some insight into how this tends to play out. There was much talk in the early stages of the pandemic about how it would catalyse a transition to a digital world. People swapped meeting up for a coffee for logging onto Zoom; bricks-and-mortar retailers closed while online shopping exploded. And naturally, the big winners were the digital giants – the sites that didn't operate in the suddenly locked-down physical world but on the internet. Leading the pack, as ever, was Amazon. Online sales of calculators, gym equipment, phone chargers, printer paper and much else besides boomed. The firm's sales grew 40 per cent in the first couple of quarters of 2020.

Amazon is an aggressively automated company, and it is easily the world's most technically sophisticated retailer. From early on, its founder, Jeff Bezos, was mindful of the power of automation. In 2002, for example, he sent a ground-breaking memo to the firm insisting that all systems in the company be designed to allow for automatic, rather than human-mediated, coordination. 'Anyone who doesn't do this will be fired,' he concluded cheerfully.[29] This edict forced Amazonians, as the retailer's employees are known, to design their internal systems to make it easier to build automated connections between them.[30] It laid the groundwork for massive automation down the line. By 2018, Amazon was able to launch physical shops that were free of any human staff. The Amazon Go stores let shoppers wander in, pick up what they want and leave, with their purchases billed automatically to their Amazon account.[31]

Amazon has also spent the last decade becoming a leader in robotics. It acquired the industry-leading robotics company Kiva Systems in 2012 for a cool $775 million. As of 2019, the firm had 200,000 robots working tirelessly around the globe, sorting billions of packages a year.[32] Amazon is possibly one of most robotised large companies in the world – with an astonishing one physical robot for every four workers. If you are one of the 200 million or so people who enjoy Amazon Prime's same-day delivery, you do so courtesy of some of those bots.

You might think, then, that the triumph of Amazon would lead to the loss of thousands of jobs. Automation, after all, is supposed to be leading to mass unemployment. Yet as Covid-19 hit in 2020, Amazon went on a hiring spree. And no small spree. In the six months after the World Health Organization declared the coronavirus outbreak a pandemic, Amazon announced four waves of hiring, amounting to a staggering 308,000 new jobs globally in one year.[33]

Amazon's example reveals that, on the level of individual companies, automation can create more jobs than it destroys. And it wasn't just Amazon. Other firms that had made significant investments in automation and AI also found themselves continuing to put people on the payroll. Netflix is a leader in AI: that, after all, is what the 'Up Next' feature is – the editorial role of determining what you might want to watch has been handed over to algorithms. But despite this, Netflix continued to hire throughout the pandemic.[34] The firm grew its workforce by nearly 9.3 per cent in 2020.[35]

Why did these supposedly super-automated firms continue to hire new people? Well, automation is usually bound up with an ambitious, rapidly growing, well-run company. When a company is expanding quickly, it needs more people – and that holds true however much you invest in automation. If we zoom out wider than the pandemic, we see that automation, growth and increasing employee numbers go hand in hand. Britain's Ocado is widely regarded as one of the most sophisticated grocery retailers in the world. It has huge, wholly automated factories where cuboid robots scuttle around picking up lettuces and ketchup and shampoo, nary a human in sight. But between 2016 and 2020, Ocado's headcount grew some 43 per cent. Chinese retailer JD.com has also invested heavily in warehouse automation. In 2018, the firm opened a warehouse that could handle 200,000 orders a day – with only four staff.[36] Yet the following year, JD.com added 48,000 employees to its payroll, an increase of more than 27 per cent.

All this means that we're left with a slightly different picture of our supposedly jobless future. The more that superstar firms like Amazon and Netflix automate, the bigger they grow; the bigger they grow, the more people they employ. There's an exponential process here, but it doesn't lead us to employee-free corporations.

Where workers do lose their jobs due to automation, it's not because they themselves are replaced by some piece of software. It's often because the firms they work for fail. And the firms they work for fail because their management or shareholders are unwilling or unable to keep up with the new possibilities of technology. That failure often extends to failing to invest in the training that their employees need to implement the latest technologies.

In other words, automation may be more like the story of two friends – call them Fred and Indrek – hiking in western Canada. They stop for a break, taking off their shoes to let their feet breathe. As they relax, they spot a grizzly bear approaching them. Fred silently starts to slip on his shoes. 'You'll never outrun that bear. Why bother putting on your shoes?' asks Indrek. 'Well,' says Fred, 'I don't have to outrun the bear. I only have to outrun you.' And then off he sprints.

If a firm is like Fred – that is, it is able to move quickly to adapt to changing circumstances – it may well thrive and grow in the age of automation, hiring more workers. However, if firms or organisations misread the nature of Exponential Age changes, their employees will get eaten by the bear. Blockbuster, the video rental service, was driven out of business by the rise of Netflix, a company originally created to send DVDs through the mail. In 2013, the last 300 Blockbuster outlets shut down, putting thousands of Blockbuster employees out of work, and Netflix emerged as the dominant player in the video rental space.[37] But those jobs were lost not because Blockbuster built some amazing automated service that reduced workforces in the 9,000 stores they operated at their peak. Rather, Netflix came up with a digitally enabled business model, first allowing customers to subscribe to DVDs from a website, and later offering them shows and films online. People preferred Netflix's offer; Blockbuster was slow to adapt and went bust.

A recent survey of 587 manufacturing companies in France supports the notion that the real threat from automation is a traditional one: the competitive threat from rival companies. The authors found that 'firms adopting robots . . . became more profitable and productive'. They also created jobs, increasing employment by 10.9 per cent. Whether the total number of production workers increased tended to depend on how fast the firm's sales grew. But in most cases, new roles would be created in other parts of the firm – resulting in a gain in employment. The problem was not so much the forward-leaning companies, but the laggards. A 10 per cent increase in robot adoption by a firm was associated with a 2.5 per cent decline in employment at its competitors.[38] It was not automation itself driving job losses, but the difficulties faced by the companies that didn't automate.

Of course, the fact that individual companies are doing well out of automation does not mean that automation isn't something to worry about more broadly. Across an economy, automation might still lead to job losses – albeit with firms like Amazon and Netflix employing increasing numbers of people.

The wider impact of automation is a point on which economists have hemmed and hawed. The French study cited above offers up a sobering picture of the macroeconomic impact of automation – concluding that, overall, automation leads to the loss of jobs across a society. The economists Daron Acemoglu and Pascual Restrepo, who co-authored the study, have conducted much of the most thought-provoking research in this area. They examined the impact of industrial robots in manufacturing, mostly in the automotive sector, and found that the machines were responsible for more than 650,000 job losses in the US between 1990 and 2007. Each individual robot displaced around 5.6 workers and reduced wages by up to half a per cent. Not so good.[39]

But today this is hardly a consistent conclusion. A European study on industrial robots, again mostly concentrated in the automotive sector, concluded that each additional robot per 1,000 workers increased overall employment by 1.3 per cent.[40] Much more encouraging. And many wider studies have suggested that, all in all, automation ends up creating jobs. In 2018, the World Economic Forum predicted automation would create 113 million new jobs over the next several years – albeit at the cost of destroying 75 million.[41] As Leslie Willocks of the London School of Economics puts it, 'as time has gone by, the estimates for net job loss from automation have been disappearing to the point of being negligible'.[42]

How can this be so? It goes against the folksy economic belief that there is only so much work to go around, and that upsetting the equilibrium of the labour force – by increasing female labour participation, or allowing immigration, or using robots – will reduce the available work for workers. But this belief is nonsense. It's a form of zero-sum thinking that has largely been dispensed with by economic theory and historical evidence. Economists call it the 'lump of labour fallacy'.

In truth, the development of new technologies also creates new needs – as one technology displaces the existing ones, new sectors of the economy are brought into existence. Those new sectors have needs that must be met by suitably skilled workers. To see how this works in practice, I'd like you to meet Sid Karunaratne. When I was building a data analytics and prediction start-up in the 2000s, we hired Sid when he was just a year or so out of university. He had the casual air of a techie – ponytailed hair and a start-late, work-late approach to his professional life. When he wasn't out rock climbing he was at his day job, tending to the cluster of servers that ran our business. These machines were the computing horsepower we needed to back up our large-scale calculations.

In his first few months, looking after this batch of powerful computers was a full-time job. Bugs in the software would cause servers to crash. Inconsistencies in the data might break our programming, causing the servers to crash. The volumes of data might fill up the storage space, causing – you guessed it – the servers to crash. Every couple of days our developers needed to update the core code that powered our business, with bug fixes, optimisations and updates for new features. Sid was responsible for putting changes into production. This meant making perhaps hundreds of little tweaks to the code, and 'pushing' it to the powerful servers that our customers accessed. And then there were all the security vulnerabilities that cyber security groups warned us of – new weaknesses discovered in the database we were using, or backdoors into the operating system. All these chinks needed to be patched up: more code that needed to be uploaded to our servers. Morning to night, Sid was busy.

As the company grew and our products improved, our computing demands increased exponentially. Our master data repository, which held perhaps a few million records when Sid joined, grew to tens of billions within a couple of years. Not only did we have to contend with thousands of times more data, we were also more demanding of it. Rather than updating the data a few times a week, we would update the repository millions of times a day, often within milliseconds of getting new information. We processed larger volumes of data, faster. This meant increasing the number of our servers – from a handful up to, at one stage, a few thousand.

But Sid wasn't overwhelmed by this more complex environment. Rather, it forced him to master new skills and build new tools. He wrote software to do key tasks for him. Soon, his job as server-sitter took up less than half a day a week. Sid figured out how to automate his own job. But that didn't lead to his redundancy. Far from it. We were a growing business, which Sid's effectiveness had helped. Automation created more opportunities, not just for Sid but for the rest of the team. With the easy tasks that comprised most of his workday now being done by code, that allowed Sid to turn to more complex, and important, tasks – many of which had previously been going unattended. And as he turned his attention to these tasks, they in turn generated more opportunities and more work.

Sid's story is not the sort that makes the headlines. But the underlying process is the one we see with many technologies, and it is playing out around the world. The futurist Paul Daugherty wrote an entire book, *Human + Machine*, about how companies that invest in artificial intelligence create new types of jobs, not just within their own company but across the economy. For example, Aurora, a start-up building self-driving trucks, is creating entirely new categories of work, which look set to become increasingly common: from the people who manage fleets of vehicles, to the remote truck operators who help them deal with unexpected problems the trucks encounter while driving.[43] In the end, across an economy automation leads to more jobs, not less.

Crucially, this dynamic only emerges in the *longue durée*. The historical record is unambiguous. Technologies have created more jobs than they have destroyed, but the short-term damage can be profound. Carl Frey points out that: 'Throughout history, the long-term benefits of new technologies to average people have been immense and indisputable. But new technologies tend to put people out of work in the short run, and what economists regard as the short run can be many years.'[44]

Overall, though, the lasting impact of automation will not be the loss of jobs. If we're looking at the long-term fallout of exponential technologies, our concern should not be with the quantity of work around for humans to do. It should be with the quality of options that are available.

At the end of 2018, when it was not quite 10 years old, Uber announced that each month more than 91 million people were using its services to order taxis or takeaways around the globe. It's a remarkable number. But a second number also stood out: that the company could count on 3.9 million drivers.[45] Not a single one of these drivers worked for Uber. They were gig workers; they didn't have an employment contract with the firm, but rather were paid, in a roundabout way, for every customer they drove – or every freelance 'gig' they completed.

Uber offers a neat demonstration of how we get work in the Exponential Age wrong. For all the talk of mass automation – talk that Uber, until recently a major investor in self-driving cars, has been all too keen to promote – the company has provided work for millions of drivers. But there is something unusual about these jobs. Uber is one of the biggest companies to use networks of freelancers, rather than contracted employees, for its primary business operations. It isn't a small company by any means – it has more than 20,000 employees, none of whom are drivers. Yet for every full-time employee, there are nearly 200 drivers – working anything from a few hours a week to 10 hours or more a day. Uber has demonstrated that platform-based gig work can work at an enormous scale. But these new working arrangements, rather than automation, are what raises the trickiest questions relating to employment in the Exponential Age.

While Uber is probably the most successful platform-based freelance work company, it did not pioneer the concept. The origins of the gig economy – where short-term, freelance tasks are allocated by an online service – lie in the Amazon Mechanical Turk platform, launched in 2005, a few years before the term 'gig working' was coined. The service gets its odd name from a famous chess-playing device of the late eighteenth century. From the 1770s, the 'Mechanical Turk' – a mannequin affixed to a chessboard that was mounted on a wooden crate – made waves by beating successive royals, aristocrats and statesmen at chess. Nominally it was powered by an ingenious machine inside the box. In fact, the Turk was operated by a human: a chess master who crouched inside the crate and manually moved the pieces.

Like the original Mechanical Turk, Amazon's version offers up an apparently automatic way to get things done. And also like the original, it is actually underpinned by hidden human labour. The interface is similar to the odd-jobs notices you sometimes find posted up on noticeboards at the back of neighbourhood stores. Tasks on MTurk tend to be quite small and well defined, but just beyond the reach of current AI systems. And so humans have to step up. A typical job, called a Human Intelligence Task (HIT), might be to go through a list of company websites, find the addresses of their branches and copy them into a database. Anyone can apply to undertake these tasks, and be paid a small fee for every one they complete. Within a couple of years of its launch, more than 100,000 Turkers, as workers on Mechanical Turk are known, were registered with the service.

At first, the tasks on Mechanical Turk had a particularly technical flavour. More than 19 out of 20 of the jobs related to getting information about digital images, or collecting information from other websites – each task being worth about 20–30 cents to the Turker.[46] The workers on Mechanical Turk became an incredible ally to companies dealing with large volumes of data. Many of the amazing machine learning systems that emerged during the late 2010s were possible because of the mind-numbing work of thousands of humans, manually classifying data for the algorithms to learn from.

In time, this type of activity garnered a new name: 'crowdsourcing'. The internet could connect people who needed something done with thousands, perhaps millions, of those with the time and the skill to do it. According to Jeff Howe, the professor of journalism who coined the term, crowdsourcing would unleash 'the latent talent of the crowd'.[47] In these early days, the notion of crowdsourcing had a utopian feel: millions of people working together, perhaps voluntarily, to build some incredible tool like Wikipedia.

Within a few years, crowdsourcing platforms had multiplied. Services like Elance and oDesk sprung up for complicated tasks like programming or copywriting; Fiverr and PeoplePerHour were created for tasks that were less complex than programming but more complex than an Amazon HIT. And if the internet got the trend started, the smartphone helped it take off. The phone became omnipresent. In-built GPS meant that phones always knew where they were – and allowing crowdsourcing platforms to offer up local, highly convenient services. Soon we could order taxis, takeaway food and massages from the comfort of our couches. TaskRabbit, now owned by furniture giant Ikea, will today dispatch someone to help you assemble your new bookcase. Talkspace will help you find a therapist. Wag! will find a walker for your dog. In time, the term 'crowdsourcing' – which often referred to unpaid, non-commercial work – gave way to a new term: the 'gig economy'.

This whole new way of working was underpinned by the emerging exponential economy. Crowdsourcing depended on digital platforms. As we've seen, these platforms were able to scale because they were susceptible to network effects and had access to an exponentially increasing amount of computing power. The tasks – sifting through data, finessing product ideas – invariably related to the development of intangible assets. And it was all facilitated by two key general purpose technologies of our age: the internet, the smartphone.

Ten years in, and gig economy platforms are continuing to grow exponentially. They have upended once-stable markets. By 2017, a mere six years after entering New York, Uber's drivers ferried more passengers than the city's yellow taxis.[48] In the US, Uber arranges more than 1 million rides every single day. Such successes turned into revenues, in 2019, of $14 billion. And that platform growth has meant more gig workers. In the same year in the UK, 2.8 million people were estimated to be platform workers – a shade under 10 per cent of those considered employed.[49] Digital platforms could add the equivalent of 72 million full-time positions to the global labour market by 2025.[50] Within two decades of the launch of Mechanical Turk, digital piecework might have increased the international workforce by as much as 2 per cent.

All this points to a truth that the hullabaloo about the robopocalypse doesn't capture. Gig work is a more imminent and transformative force than mass automation. But what does it actually mean for workers? Evangelists for the gig economy – among them, naturally, the founders of leading gig companies – say that their model helps workers in two key ways: it can make markets bigger and more efficient, creating more opportunities for workers; and it can improve the quality of work that someone does.

Many labour markets are inefficient. Demand for a particular type of work goes unmet, perhaps because it is hard for employers to find workers or vice versa. Or perhaps there are middlemen taking an unfairly large piece of the pie. Gig-working platforms make it easier to connect buyers and sellers – the databases and algorithms do the matching. This can lead to an overall increase in opportunities for workers. In developed economies, Uber is bigger than the taxi businesses in many big cities – evidence that the company is expanding markets. In emerging economies, labour markets are often clunkier. Kobo360, a kind of Uber for freight, has helped Nigerian truckers get work in a famously inefficient market mired in corruption and bureaucracy.[51]

One key way platforms make markets more efficient is by making it easier to discover and apply for jobs on the other side of the world – which might otherwise have been hard to fill. While a delivery driver or masseur is constrained to where they live, designers, programmers and copywriters can work wherever they have an internet connection and share a language with a client. The platform UpWork is a great example of how the gig economy can help workers access the global economy, often in part-time roles. I've personally used the platform to hire developers from Egypt, Bulgaria, Pakistan and Colombia, a sound editor from Croatia, and designers from India – all in part-time roles. In 2011,

UpWork freelancers billed about $200 million in jobs like this on the site. By 2020, that number had increased more than 12-fold.[52]

The second putative benefit of gig working is that it might qualitatively improve the nature of work. In the developed world, what the pro-gig-work crowd normally focuses on is flexibility. A gig worker can work when they want rather than under the permanent subordination of an employment contract. Lyft and Uber drivers show remarkable levels of satisfaction with the flexible work set-up: 71 per cent of drivers want to remain independent contractors, albeit down from 81 per cent prior to the Covid-19 pandemic. When asked what is most important to them, drivers ranked pay and flexible schedule as their top priorities.[53] In a similar vein, a 2018 British government survey reckoned that more than half of gig workers were satisfied with the independence and flexibility provided by their jobs.[54]

If gig work is generally more flexible and less formal in richer countries, the reverse is often true in poorer ones. In emerging economies, a gig-working platform may offer more security, more employment options and greater freedoms than casual or day labour does. In India, for example, the sheer size of the informal labour market gets in the way of the government being able to spend on healthcare and education. Casual labourers, hired daily and paid in cash, rarely pay income tax. Nor do their employers contribute to payroll tax. Lower tax participation means less booty in government coffers to fund those social programmes. For highly casual labour markets, the gig economy could be a route to a large, more formal sector – with more protections for workers and a more robust tax base for governments.[55]

So far so good. Yet there is evidence that all is not rosy for workers on digital labour platforms, especially in advanced economies. Pay is often poor compared to traditional work; working patterns can be precarious, offering few protections should a worker get sick. The platforms themselves can make unilateral changes about how they operate and what they pay. Many companies maintain internal scores about workers on their platforms, which might affect what jobs they are offered. And equally, unions or other collective arrangements are uncommon among the independent workers on gig platforms – which means there is often no collective voice to represent their interests. What this boils down to is a huge imbalance in bargaining power between the platforms and their armies of labour.

One way this power imbalance manifests is through low pay. The British government's survey from 2018 concluded that nearly two-fifths of gig workers earned less than the equivalent of £8.44 per hour, only a little over the British national minimum wage.[56] Low pay in the gig economy is also common in the EU. Platform workers in Germany make 29 per cent less than the local statutory minimum wage, while workers in France make 54 per cent less than its minimum wage.[57] Because gig workers often have to meet their own expenses – repairs, fuel, even uniforms – take-home pay can get squeezed. One study from the Massachusetts Institute of Technology estimated that the average Uber driver makes a profit of $3.37 per hour. In 2019, researchers found that American workers for DoorDash, a food delivery service, made $1.45 an hour after expenses, or about a fifth of the then national minimum federal wage.[58] (If you are using a gig-based delivery service, it might be worth tipping handsomely.)

As these smartphone-based gig companies sprung up in the mid-2010s, they also led to increasingly precarious working conditions. The new tech platforms go to great pains to explain that the people doing the work are not their employees. The platform's job is merely to introduce you to someone who will drive you to work or deliver you some late-night ice cream. They are like a modern-day temping agency.

Such distinctions matter. In many countries, particularly in the developed world, employees are treated markedly differently to the self-employed. Employment involves a clear trade-off. The employee enjoys a bundle of benefits, such as a regular wage and job stability, as well as hard-won labour rights: the right to an annual wage, sick pay, parental leave and a fair dismissal. In return, the employer gets the time and best efforts of the employee. The deal is stability, in exchange for subordination. The self-employed, on the other hand, have always been at the mercy of the market. Work can be irregular, and sickness means days off work with no pay.

Companies using gig labour, the largest of which are in the ride-hailing and food delivery sectors, have generally been reluctant to strengthen the protections they offer their workforces. The state of California passed Assembly Bill 5 in 2019, which mandated freelancers to be classified as employees and so have access to the requisite perks. Uber, Lyft and DoorDash were not keen. Through the most expensive lobbying effort in Californian history, they successfully got the state to pass Proposition 22, which granted them an exception – to keep classifying their drivers as independent contractors, albeit with some wage and health protections.[59]

A similar battle was underway across the Atlantic. In London, James Farrar and Yaseen Aslam, two drivers, took Uber to court, arguing that they weren't self-employed but should be considered workers under British law. (The classification of 'worker' in British law is a curious one – it does not offer all the protections of an employee, but more than those of a freelancer.) But this time the outcome was different. Uber fought Aslam and Farrer for five years, until finally the UK's highest court dismissed the taxi platform's arguments unanimously.[60] Arriving at the decision, the judges concluded that the drivers were not self-employed entrepreneurs in any sense of the word. Uber controlled all aspects of their work, including the number of trips they received, the price of their service, and even the communication between driver and passenger. Lord Leggatt, who wrote the ruling handed down by the Supreme Court, wrote: 'The question . . . is not whether the system of control operated by Uber is in its commercial interests, but whether it places drivers in a position of subordination to Uber. It plainly does.'[61]

In part, these teething problems are down to an exponential gap – between new modes of employment enabled by exponential technologies, and a set of labour laws designed in the twentieth century. The great triumphs of the twentieth-century labour movement were about securing humane working conditions for contracted employees. The eight-hour day, sick leave, pensions, collective bargaining – all were extended to those who were formally employed by a company. But in the Exponential Age, formal employees are relatively decreasing in number. The technologies of this era create new ways of working, with the smartphone and the task-matching algorithm allowing firms to rely on pools of freelance talent. And our labour laws haven't yet caught up. That workers are forced to rely on court decisions – rather than clear rules – reveals that the gap is far from closing.

All of this leads to a growing inequality between gig workers and official employees. The self-employed have always had to face the whims of their clients. But in the Exponential Age, their number could swell to the hundreds of millions. Only a small group will retain the privileges that workers have fought to gain over the last 150 years.

Willard Legrand Bundy was an inventor of the old school. Impeccably moustachioed and besuited, Bundy is usually pictured standing proudly beside one of his myriad inventions – a calendar-clock, adding machines, cash registers and more. But one of his simplest devices had the greatest impact of all. His 'time recorder', a device that allowed employers to track when their workers clocked in and out, became one of the defining inventions of the twentieth-century workplace. Time recorders first appeared in the 1880s, and by the turn of the century close to 9,000 had been produced. Employers finally had an automatic way of keeping tabs on their workers.

Bundy's invention represented an early move towards a more scientific, regulated, empirical system of people management. Since Bundy's time – he died in 1907 – methods of managing employees became progressively more sophisticated. Methods like Bundy's would be applied systematically across industries.

The tipping point was the rise of scientific management, pioneered by Frederick Winslow Taylor. Like Mark Zuckerberg more than a century later, Taylor attended the prestigious boarding school Phillips Exeter Academy before attending – and dropping out of – Harvard. He went on to work in a machine shop in the steel industry at the age of 19. During his experiences on the shop floor, he noted that workers soldiered at a 'slow easy gait' because of a 'natural laziness'. Through close observation – including what became known as 'time-and-motion studies' – he started to quantify human behaviour at work and use his insights to develop rules to optimise output. To boost productivity, he preached, managers should measure everything, link pay and bonuses to performance, and break complex tasks into simpler ones.[62]

Taylorism underpinned much of the productivity growth of twentieth-century businesses. But it was also punitive, leading to constant surveillance of workers, draconian punishments for 'underperforming' employees, and a general tendency to treat workers like machines. The rule of Taylorism is that the unobserved worker is an inefficient worker.[63] And as time went on, Taylorism would only become more intrusive in the lives of workers. In the 1930s, Lillian Gilbreth – a psychologist and engineer – developed personality and psychological testing for personnel management staff (who would now be known as 'human resources'). This approach would soon become the norm in large organisations.[64] As one contemporary account of workplace surveillance put it, 'Not only would workplaces . . . be designed so that workers would internalize their boss' gaze, but the addition of these testing methods signalled that the boss was genuinely trying to get inside employees' heads.'[65] It was that classic twentieth-century deal: you get a high-security, high-wage job, but in return you give up your autonomy.

However, by the turn of the millennium, some workplace norms were shifting. Among technology companies, Taylorism was slowly abandoned, replaced by greater freedom for employees. The exponential revolution seemed, at first, to catalyse the long-term shift to a less punitive workplace culture. It was in the run-up to the dot-com boom that companies started to care about their employees. Gone were grey cubicles, vending machines and fluorescent lights. In came bean bags, free lunches and craft beer. Among tech firms, fun amenities became the norm: pool tables, sleeping pods, table football and free food. In 2000, I sat on the board of an internet company that featured a real lawn and swings in its headquarters on the second-floor of a former warehouse in North London.

By the second decade of the twenty-first century, this was fast becoming the norm. Asana, the firm behind a popular team management app, provides its employees with organic home-cooked meals, life coaching and yoga classes. Another software company, Twilio, gives unlimited time off and offers free on-site massages twice a month. By 2017, oocyte freezing – a procedure young women can undergo to freeze their eggs, to delay having children – had become a perk some technology firms provided to their employees. And not only technology firms: Goldman Sachs offers to cover its employees' fertility treatments and egg freezing up to $20,000.

Sometimes it seemed that the bigger the superstar company, the more generous the offer. From 2009, Netflix gave its employees as much vacation time as they wanted, provided they got their work done. Google, meanwhile, was a pioneer of unusually generous death benefits – if a Google employee dies, the company gives their surviving partner 50 per cent of their salary for a decade after their death – which is unusual even by Exponential Age standards.

Even before the Covid-19 pandemic, many companies were used to employees working remotely at hours that suited them. In the first firm I founded, in 1999, Wi-Fi was a rarity, and public Wi-Fi was non-existent. Most people were largely confined to working from the office, their computer tethered to a network socket – perhaps sometimes taking a laptop home to do a bit of digital paperwork. By the time I founded my second start-up a decade later, the tools of digital work were much richer. We've already met Sid Karunaratne, with his penchant for automating his own day job. The tools that allowed Sid and his team members to work from home (or the beach) were wide-ranging – and where possible they took advantage. This kind of unregulated, arm's-length management is as far away from Taylorism as one could imagine. And it only accelerated during the coronavirus lockdowns, when internet-enabled remote work became the norm for most white-collar workers.

But not every employee works for a firm which provides the freedom to work how and where they like. And the same technologies that enable remote work – from a beach, a mountain or by a lake – can also be turned against employees.

Many readers might be familiar with the 'panopticon'. Imagined by the early-nineteenth-century philosopher Jeremy Bentham, the panopticon is a type of prison in which every prisoner is constantly visible from a guard tower in the middle of the complex. In such an institution, prisoners never know if they are being watched – because they could be under surveillance at any time. And so, the theory goes, their good behaviour is guaranteed.

Today's workplace devices are a little like an inescapable digital panopticon. Because every instant message, every email and every document update is logged, you never know when you are being watched. And so, perhaps, you have no choice but to behave.

These methods of surveillance are increasingly common. Many firms today use facial recognition and mood detection systems, or are developing software that estimates how engaged and satisfied employees are. These devices scan emails, Slack messages and even facial expressions for hints of a worker's mood. The Japanese tech giant Hitachi created name tags with sensors, designed to measure who employees speak to and for how long, to track and measure their movements, and to feed the data into the 'organisational happiness' index.[66] In November 2020, Microsoft updated its Office software – the most widely used of its kind in the world – to include 'Workplace Productivity' scores which monitor how employees use everyday apps like Word.[67]

And as the technology gets more advanced, it gets more invasive. Some factories in China have garnered headlines for introducing brainwave-reading hats to track their workers' emotions and focus so they can adjust the length of break times to reduce fatigue.[68] Increasingly, automated monitoring systems are used even before a worker joins a company, in the recruitment process. Unilever used AI to save '100,000 hours of interviewing time and roughly $1m in recruiting costs each year' by delegating job interviews to video analysis software.[69]

Today, these technologies are generally reserved for the most tech-savvy companies. But, as so often in the Exponential Age, there is only a few years' gap between the technological vanguard and the rest of us. And the process is further along than you might think. A 2018 Gartner report found that half of 239 large corporations were monitoring the content of employee emails and social media accounts.

The move to remote work during Covid-19 only accelerated this transition, particularly in white-collar sectors that were previously office-based. A report by the UK's human resources trade body, CIPD, found that 45 per cent of employees believe that monitoring is currently taking place in their workplace.[70]

The issue is not just surveillance, but wider forms of automated management. Workers on platform apps have diminishing control over how they work. When dozens of gig workers for Uber Eats gathered outside the company's office in south London in 2016 to protest, they were not only criticising their low pay. They were questioning the core of what makes the gig economy giants, especially ride-hailing companies, successful: algorithmic management. 'We are people, not Uber's tools,' they yelled.

These people and millions of other gig workers are managed by computers. Their work is scrutinised through a stream of quantitative performance assessments. Rideshare drivers may only have 10–20 seconds to respond to an offered ride, without knowing in advance where they're expected to go or how much they can expect to make. If they refuse too many rides in a row, they can be kicked off the platform.[71] And this is not limited to ride-sharing. The warehouse workers who put together your grocery delivery are commanded by scanners, which tell them how much time they have to collect each item.[72] None of Taylor's lackadaisical soldiering here. Deliveroo drivers, meanwhile, can be punished if they arrive later than the time of arrival calculated by an algorithm – regardless of the local weather conditions or traffic.

This system of management works like Taylorism on steroids – with all the dehumanising downsides but without the commensurate high pay or job stability. One Amazon worker's claim, reported in *The Guardian*, that the company leaders 'care more about the robots than they care about the employees' encapsulates one of the oldest criticisms of scientific management.[73] Except, this time round, workers don't have the collective bargaining power they once did. In the ultra-Taylorist companies of the twenty-first century, the power balance is skewed heavily in the favour of bosses.

Amazon automatically fires about 10 per cent of its factory staff annually, for not being able to move packages through the system quickly enough.[74] The company's 'proprietary productivity metric', which dictates how quickly workers must process each order, gets our one-day Prime deliveries to us on time. But it also locks workers out of important aspects of decision-making, and denies their agency. One liquid package bursting open on a conveyor belt will take minutes off the targets set by the algorithm – hundreds of packages per hour – becoming one more black mark on your way to getting fired. And package bursts are not rare occurrences.

All this seems paradoxical. In many cases, the very same companies who offer select employees ping-pong tables, craft beer and egg-freezing treatments are obsessively monitoring and controlling others via algorithms. This is the Janus face of work in the Exponential Age. Those who are well-educated and lucky can thrive. Those who aren't might find themselves trapped in an increasingly punitive workplace.

You may have noticed a pattern emerging. The future of work seems less defined by the absence of work and more by a growing chasm – between increasingly high-quality work for some, and increasingly low-quality, insecure work for others. We aren't on the verge of a jobless future, but we are perhaps looking at a future in which – if we aren't careful – work no longer serves the interests of society.

This problem is visible in the precarity of gig work, and the treatment of employees as fungible assets to be controlled through management algorithms. But its clearest manifestation lies in how the rewards are doled out. Economists have long tried to understand how equitable a labour market is by looking at what share of a country's income goes to workers, and what share goes to the owners of capital – through stock gains, dividends and corporate profits. The story of the last 50 years is striking. Between 1980 and 2014, labour's share of national income – the percentage of GDP paid out in wages, salaries and benefits – declined on average by 6.5 per cent, as measured in 34 advanced economies. In the US, the decline has been even more staggering. In 1947, workers received 65 per cent of the national income in America; by 2018, this had declined to 56.7 per cent. More than three-quarters of the decline since the end of World War Two occurred in the first two decades of the millennium. For decades, workers have been systematically getting a smaller share of the economic pie.[75]

In practical terms this has manifested through stagnating average wages and increases in inequality. Again, the most striking example comes from the US. Between the 1940s and the mid-1970s, economic productivity and workers' pay rose in tandem: from 1948–1973, there was a 97 per cent increase in workers' hourly pay, against a 91 per cent increase in economic productivity. Fair enough. But then, something surprising happened. The increase in wages tapered off – even as economic productivity continued to skyrocket. By 2018, US economic productivity was 255 per cent higher than it had been in 1948; but workers' pay was only 125 per cent higher – barely a third more than it had been in 1973. In other words, the output of the US economy continued to rise – but the share workers received stagnated.[76]

The cause of the decline of labour's share of the economic pie is multifaceted. But it is closely related to the shift to the exponential economy. Four key causes stand out. Globalisation, which drove down wages in the West as companies offshored jobs to cheaper locations across the world. The decline of unions, which meant workers lacked the bargaining power to stop economic rewards going to the owners of capital (we'll return to this in a moment). The rise of the intangible economy, which reduced the relative value-add of the average worker – more value was being created by know-how, software and data, stewarded by smaller numbers of highly specialist workers, than by the human sweat of the larger parts of the workforce. And superstarification – as markets consolidated around ever-fewer superstar firms, there was less competition for labour and so workers had less leverage.[77] These are all hallmarks of the rise of exponential technologies – the emergence of a global, high-tech, intangible economy, dominated by a handful of big firms. According to one study, more than two-thirds of the loss to workers results from the transition to intangibles and the growth of superstar firms.[78]

Of course, not every job created by exponential technologies sees wage pressure. Take a company like Uber. It depends on highly sophisticated software: from the algorithm that allocates a driver, to the machine learning that helps forecast demand, suggest destinations, predict bottlenecks and establish surge pricing. All of this software has been designed by engineers. Getting an engineering job at Uber requires a strong academic background as well as top-tier experience working on the latest technologies. The selection process involves several hours of interviews, as well as coding tests and advanced problem-solving – the kind of rigorous testing process that is a hallmark of superstar companies. Should you get the job, though, it all becomes worth it. The average software engineer at Uber was paid $147,603 a year in 2020. Senior engineers with five or more years of experience might make three times that. On the flip side, there are Uber's drivers. As we've seen, gig workers are often paid relatively little. Your typical driver will make $19.73 per hour before expenses, or $30,390 a year if they drive 40 hours a week.

There is a similar dynamic at work in Facebook. Half of all employees at Facebook, from engineers to marketers, accountants to salespeople, make $240,000 a year or more.[79] Facebook's content moderators, who are not employed by the firm but rather contracted via temping agencies, are paid on average $28,000 per annum. Ordinary people who create the contacts, content and conversation on Facebook are, of course, paid nothing.

And this is not just true of digital platforms. Zymergen, a breakthrough company working on the intersection of biology and artificial intelligence, has a similar bifurcation of its workers. On the one hand, highly paid scientists with doctorates; on the other, lower-paid support staff – and little demand for mid-pay workers between these two extremes.[80]

All this points to the changing topology of employment in the Exponential Age. This is an economy where intangible assets – the kind produced by well-educated knowledge-workers – are all-important. Those with high levels of education are compensated handsomely. At the same time, there remains a group of less well-rewarded, less highly skilled workers, who may not even be acknowledged as employees. In aggregate, the result is a reduced share of income that goes to employees. Middle-wage earners, who used to be the engine of Western economies, are evaporating.

In the long run, workers have generally done well from new technology. It has normally led to innovations that improve working life. The plough, electricity and indoor lighting all made labour a less hazardous task for many workers. For all the tumult of the industrial revolution, it did eventually lead to a marked, consistent and sustained increase in the living standards of workers.

But global labour markets are complicated. There are winners and losers. In Britain, as wages rose sharply through the nineteenth century, some groups were immiserated – the famous handloom weavers, forced into obsolescence by the invention of the mechanical loom, for example.[81]

Even as salaries rose, many workers had to toil in the hellish conditions we met in Chapter 3. Between 1790 and 1840, wages of workers rose by a modest 12 per cent, while GDP per worker increased by more than half. It was not until the 1860s that ordinary people's wages caught up with the gains of technology; and not until 1900 that they levelled up entirely. There was a century where workers lagged behind the gains; an uncomfortable century, even if it worked out in the end.[82]

'Worked out in the end' are five very easy words to write. But they represent decades of catch-up – papering over lives where incomes didn't grow, and quality of life fell relative to the strength of the economy. As John Maynard Keynes put it, 'In the long run we are all dead.'[83] In the meantime, there will be pernicious chasms in the labour market, between the changes wrought by new technology and the norms and laws that govern the workplace.

But it is not too late to redesign the economy to close this widening gap. As so often in the Exponential Age, even as disruptive new technologies take off, remoulding our economy and society in the process, so too do new ideas about how we might adapt. It just takes a little while for the ideas to become a reality. Today, we can already make out the four characteristics we need for a fairer work settlement.

To start with, workers need dignity. Many of the changes described in this chapter have had a dehumanising effect on workers – thanks to the power of Exponential Age technologies to observe, monitor, rank and cajole employees. It's interesting, however, that these technologies don't need to be used in such an ominous manner. For example, BP America use Fitbit bracelets to monitor employee wellbeing – a method that gently encourages staff to do more exercise, something for which many workers are grateful. Crucially, this scheme is voluntary. But it shows that new technologies – even surveillance technologies – don't inherently lead to exploitation. Provided workers are given a say in how these technologies are used – and are given the opportunity to reject them – new workplace technologies need pose no threat to workers' wellbeing. In fact, they could offer a path towards greater employee satisfaction.

But to make a dignified workplace possible, we need to change our approach to management. That means finding new ways to prevent bosses taking advantage of unfair power differentials – with employees being given greater autonomy over how they work. For workers increasingly managed by algorithmic systems, that might mean getting a chance to speak to a real person to seek redress when the algorithm has been unfair. Workers should also be given more insight into the data that firms hold about them, and clearer explanations on how automated decisions are made. Unions are increasingly mindful of how they can help workers win back 'data rights' about who gets given tasks, who gets fined and who gets monitored.[84]

Flexibility also matters. Work in the Exponential Age is volatile. Ours is a time of dramatic transformations. Firms will fail. Old industries will hollow out. Entire technologies, like fossil fuels, will become redundant. At the same time, new technologies will continue to emerge and grow– cheap solar power, quantum computing, long-term battery storage, precision biology. These technologies will lead to the emergence of dynamic new industries, which will require more workers with different skills. This constant change is what has driven many of the problems identified in this chapter: legacy firms going bust, with their employees left unsuited to the new roles that are being created.

To adapt to this new age, workers need to be given the chance to constantly reskill. In the old world, you might have a single job from the ages of 18 to 60. Today, few industries look likely to stay around for that long – and if they do, it will be through constant technological upgrades. The World Economic Forum expects 42 per cent of the core skills needed for existing jobs to change between 2020 and 2030. But how can we make constant reskilling possible?

Digital technology can help us here too. It seems fatuous to point to TikTok as an example of how people can reskill – but apps like this may well be the future of education. TikTok is rife with DIY videos – particularly in lifestyle areas like home improvement, cooking and skincare. But people with specialist interests can also find communities and continue to learn. Teachers, academics and nutritionists, for example, all have big communities on TikTok, and many use their followings to disseminate information or to correct widespread misconceptions. In June 2020, TikTok announced that it would be commissioning experts and institutions to produce educational content as part of a new trend for micro-learning. Of course, TikTok alone will not reskill the economy – but it points to nascent forms of digital learning, the potential of which is only now being explored. Why shouldn't we all be constantly educating ourselves via digital platforms?

Workers also need security. In times of rapid change, a safety net becomes critical – lest people lose their jobs and find themselves unable to survive. And the more entrepreneurial and volatile the economy, the more essential such a safety net becomes. Many academics and technologists, from the French superstar economist Thomas Piketty to the founder of the Web, Tim Berners-Lee, argue for universal basic income (UBI) to solve this very problem. Under a UBI system, a government gives every citizen a regular sum of cash, no strings attached. Expensive as this might sound, it's certainly a quick route to economic security for large numbers of people who might otherwise be at the mercy of a cruel labour market. Small-scale experiments in the US city of Stockton and Finland's capital Helsinki have shown that people receiving UBI report increased wellbeing and lower levels of food distress. And when they lost their jobs, UBI recipients were nearly a third more likely to get back into work within a year, relative to those who were on more traditional forms of assistance.[85]

But UBI, at the moment, seems a distant possibility. It remains plagued by concerns that it may weaken work ethic and create a culture of dependency. While early studies suggest this isn't the case, in many countries UBI may be too controversial to be a viable policy for the time being. A cheaper, more politically palatable alternative is already in place in Denmark – and is known as 'flexicurity'. The flexicurity model has two sides. On the one hand, employers can hire, fire and tweak employment terms at will. On the other, employees are guaranteed extensive protections and benefits from the state should they find themselves unemployed. For example, an unemployed Dane can make up to 80 per cent of their previous salary – and more if they have kids – provided they can demonstrate they are looking for work.

This would solve one of the great problems of the Exponential Age. Firms don't necessarily know how fast they will grow, how quickly a business opportunity will emerge, and which blind alleys they will wander down – so flexibility in hiring and firing is valuable. But for workers, of course, agility for corporations doesn't pay the rent. The Danish model gives firms the adaptability they need, and the workers the security they need.

Finally, workers need equity. While labour's share of the national cake has been declining for decades, new technologies look set to accelerate this process even further. Technology, globalisation, the intangible economy – all shift power away from labour and towards capital. As the Exponential Age speeds up, we must ask ourselves: How much economic reward should go to entrepreneurs, owners and workers? And is it time for workers to be given a bigger share?

If the answer to the latter question is 'yes', there are a few changes we need to make. For one thing, employers need to invest much more in their workers' development – so they can move into more skilled, better-remunerated roles as technology develops. But this may accentuate the gap between the pay cheques of low- and high-skilled workers. Ours is an era when a programmer for Uber makes at least four times more than the average driver; and in which evidence from the UK shows that for every 10 new high-tech jobs, 7 new low-wage service jobs are created.[86]

And so, in the Exponential Age, minimum-wage bargaining becomes more important than ever, and policymakers need to make sure it actually reflects the cost of living. At the same time, governments might explore how to mitigate the growing wage differentials between the richest and the poorest workers. One way to make this more likely would be through greater wage transparency. Firms already disclose how much they pay their top directors, and increasingly share data on the gender pay gap. Encouraging companies to share more information about their wage practices would illuminate the problem of income disparity.

This perhaps all sounds utopian. And it's true: closing the clutch of exponential gaps in the workplace will take great effort. But it is far from impossible. In fact, there are a number of organisations who are already working to close those gaps. They hint at how, in practice, we might achieve the legislative and economic shifts we need.

In the absence of quick action by lawmakers, some entrepreneurial outfits are devising ways to directly support workers. Gig workers often struggle to get access to loans and mortgages; lenders look to strong credit records, and the agencies that offer credit ratings prefer those on standard employment contracts. Portify, based in London, is a start-up that helps freelance workers build up their credit ratings month by month. It effectively sets up a tiny loan for members, which is paid at a rate of £5 ($8) a month. Credit reference agencies like consumers who take out loans, and they love those who pay their loans on time. A positive repayment history makes up more than a third of your credit score. Portify offers just such a history for its gig-working members.

But entrepreneurship alone will not be enough. History shows that meaningful improvements for workers rarely come thanks to the benevolence of bosses, but from pressure by employees. One reading of nineteenth- and early-twentieth-century history is as a long struggle by labour to organise. In the UK, the modern labour movement arguably began in the 1880s, with the emergence of 'new unionism' – a wave of trade unions that explicitly appealed to the urban, often-unskilled working class. In the US, the 1937 Battle of the Overpass – when angry Ford workers who were determined to form a union were beaten by Henry Ford's security entourage – was a decisive moment, after which Ford agreed to sign a contract with the United Autoworkers Union and raise workers' salaries. The modern social democratic parties of much of the developed world are the legacy of these struggles and others like them.

But unions have fallen out of fashion in the Anglosphere, following a 50-year assault by the political establishment. Union membership in the

UK, US, France and others is at its lowest level since records began.[87] The free-market orthodoxy we met in Chapter 1 had little time for trade unions – they were seen as 'labour cartels' that illicitly interfered with the workings of the market. In the UK, Margaret Thatcher's government perceived trade unions as one of the biggest internal threats, and took steps to significantly weaken their power. Between 1979 and 1988, union membership declined 20 per cent, thanks to government policies – combined with economic hardship and a decline in manufacturing employment. The US was on a similar trajectory. In 1981, the newly elected Ronald Reagan fired over 11,000 striking air-traffic controllers and replaced them with non-union members. It marked a turning point in the history of American trade unionism.

As a result, there was little unionisation in the early tech industry. This was still true as we entered the Exponential Age: the workforces of the digital superstars were completely un-unionised. The tech industry has been anti-union since its inception – Robert Noyce, who co-founded Intel, declared that 'remaining non-union is essential for survival for most of our companies'.[88] Today's biggest companies have arguably picked up this mantle. American senators Elizabeth Warren and Bernie Sanders have long accused Amazon of heavy-handed attacks on workers who try to unionise.[89] Throughout the last decade, workers at tech companies who try to organise have run into many obstacles – in some cases because they didn't get the widespread support that they needed, and in others because of active interference by their employers. Only in January 2021 did roughly 200 workers from Google's owner, Alphabet, finally form a union – but out of an employee pool of 135,000.

Perhaps the word 'union' just has too much baggage in the Exponential Age. It conjures up images of striking miners and steelworkers more than of striking Deliveroo drivers. Yet collective action remains the best way to guarantee a workplace that is fair for employees. If we are to build an employment settlement that is dignified, flexible, secure and equitable, workers will need to get organised. Only unions can collectively bargain on behalf of workers. And unions are also better than individual workers at developing the expertise required to digest the complex technical issues emerging in a rapidly changing economy.

But in the world of trade unions, too, digital technology is both the problem and the solution. The creation of new networks of information gives workers an opportunity to identify their shared experiences, discuss how to respond, and organise. Finding like-minded people to collaborate with has always been a barrier to unionisation – but this problem is less pronounced in the Exponential Age when anyone with a smartphone becomes a potential comrade. Today, the process of unionisation often begins with informal collectivisation on WhatsApp groups and online forums. While workers in each industry have their own demands, the use of technology for collectivisation is uniform.[90] In West Virginia, where union membership rates hover at around 10.5 per cent,[91] teachers looking to organise set up an invite-only Facebook group. Close to 70 per cent of the state's 35,000 teachers joined, turning the group into a hub for discussion and coordination of major statewide strikes.[92]

When organised online, such collectives can be truly global in a way internationalist trade unionists of the early twentieth century could only have dreamed of. In January 2020, gig-economy drivers from around the world gathered in Oxford for the first meeting of the International Alliance of App-Based Transport Workers, which was organising for better working conditions for all app-based transport workers globally.

These digitally enabled unions can close the exponential gap by making sure that employment laws and worker-employer relationships are constantly adapting as the economy changes. Collective bargaining and worker organisation have long been the best ways of ensuring that employers remain conscious of the needs of their workers – and that laws follow suit. Without them, workers may well find themselves unable to keep up with the clip of the Exponential Age.

The great risk facing workers in the Exponential Age doesn't come from robots. It comes from a rapidly changing economy – defined by a fundamental shift in the quality of working arrangements, resulting from gig-working and algorithmic management. For workers, this leads to an age-old problem: a power imbalance between bosses and workers. But while this imbalance is a consequence of the Exponential Age, it is not an inevitable one.

6

The World Is Spiky

Angelo Yu had a problem. It was late 2019, and US president Donald Trump had spent much of the previous two years tweeting increasingly bellicose denunciations of the Chinese government. At the same time, the White House had been progressively ratcheting up tariffs on Chinese imports to the US. This looked like bad news for Yu's start-up, Pix Moving. Based in Guiyang, a city 1,000 kilometres to the north-west of Shenzhen, the firm makes the chassis for a new class of autonomous vehicle. The tariffs made everything more expensive.

A lesser entrepreneur may have had to raise prices for his first customers. Not so Yu. He had a solution: Pix Moving was using only the most modern manufacturing methods – 'dematerialised' techniques. Rather than exporting cars, Yu explained, they 'export the technique that is needed to produce the cars'.[1] Vehicles are not loaded onto container ships and sent to their destination. Rather, the company sends design blueprints over to colleagues in the US, who use additive manufacturing techniques to print components locally. From those components, the finished product can be assembled. Yu's approach could skirt around customs inspectors (and tariffs). Additive manufacturing lets him build wherever his customers are, trade conflicts be damned.

Pix Moving is a start-up that reflects how manufacturing will change as we move further into the Exponential Age and, in doing so, unpick our assumptions about globalisation. On one level, Yu's story reveals a highly globalised economy in action: a car can be designed in Guiyang and assembled in California with remarkable ease. But it also represents an inversion of globalisation – a return to the local. For decades, supply chains have been getting longer; production processes more international. The various components of a car might be manufactured in a dozen countries and assembled in more than one. In the future, however, manufacturing can happen near to the consumer. Thanks to the wonders of 3D printing, components can be produced locally; while the design might come from anywhere, the finished product can be crafted in a local workshop and handed to a customer who lives close by.

It wasn't meant to be this way. Thomas Friedman's bestselling 2005 book *The World is Flat* declared itself to be a history of the twenty-first century. Its argument: the world is entering a third phase of globalisation. The first, which began with European exploration of the Americas, is most frequently associated with colonialism and the globalisation of trade between countries. In the second, which got going in the nineteenth century, the focus shifted to the activities of transnational corporations – culminating in the monolithic industrial firms of the post-war era. In Friedman's third phase, globalisation would reach a new level – with flows of trade, labour and information becoming ever more international. Friedman identifies ten 'flatteners' – from workflow software and outsourcing, to the development and expansion of sophisticated supply chains – each of which will make the world more globalised.

Friedman was one of the great prophets of globalisation, but he wasn't the first. By the final third of the twentieth century, there was an array of international institutions propping up a globalising economic order: the World Trade Organization and the Organisation for Economic Co-operation and Development; the International Monetary Fund and the World Bank. Multilateral institutions, including the European Union, were helping flatten the world too – creating multinational political systems premised on free internal trade. The private sector had its own infrastructure of globalisation. Annually at the World Economic Forum in Davos, political and business elites would get together to find common ground in this increasingly flat world.

The power of globalisation was transformative. In 1970, trade represented about a quarter of global GDP. By 2019 it comprised nearly 60 per cent of a much larger global GDP. This brought many benefits. As international trade grew, other forms of global exchange became more straightforward and appealing, from holidays abroad to students studying overseas. Above all, globalisation coincided with a huge increase in wealth around the world, as markets expanded and nations were able to reap the rewards of more open trade. According to its evangelists, globalisation has contributed to better standards of living and the reduction of poverty across the planet.[2] As I argued in Chapter 2, the growth of globalisation created very large markets that helped speed up the development of exponential technologies – and the growth of the new businesses that took advantage of them.

Yet following the global financial crisis of 2007–2009, globalisation started to lose its lustre. It had grown in tandem with the financialisation of national and global economies – as trade grew, so too did the importance of borrowing and lending, often through increasingly complex financial instruments. When the financial crisis hit, the pain was not limited to investors – it spread into the 'real' economy. In richer countries, many felt that globalisation had led to the offshoring of blue-collar jobs to the emerging economies of the developing world. After 2010, there was an increasing turn towards nationalism in many countries: the Brexit vote in the UK, and the election of Donald Trump in the US. While globalisation remains a potent force in the world economy, it is also increasingly unfashionable.

This is a story that most readers will be familiar with. Much has been written about how globalisation lost its appeal. These accounts usually point towards growing income inequality, concerns about immigration, and deindustrialisation in developed nations – all driving a turn against the supposed benefits of a flat world. Less well-scrutinised, however, is the way exponential technologies both create the rationale for more borders and provide the tools to build them.

We often assume that the more high-tech a society becomes, the more global and borderless it will be. And until recently, that has often been true. But not any more. Many exponential technologies lead to a return to the local. These breakthrough technologies favour the near over the far. It's not just the 3D printers relied upon by entrepreneurs like Angelo Yu. Exponential technologies also facilitate the local production of energy and food, in a way that would have been prohibitively expensive until recently. And new technologies, and the businesses built on them, often need large numbers of people interacting with each other in close proximity – something only cities can offer.

As the twenty-first century unfolds, the localising potential of technology will only become more powerful. The coronavirus pandemic which began in 2020 showed how fragile global supply chains could be. But if it was a virus in 2020, it could be war or extreme weather – exacerbated by anthropogenic climate change – in the future. The result is an era in which, once again, geography matters – with economic activity set to become increasingly local.

There is an irony here. The economic paradigm that brought about the Exponential Age, globalisation, has fostered technologies that will lead to a return to the local. But our political and economic systems were not designed to cope with the new age of localism. As so often, gaps emerge. Between the economic policies advocated by our political institutions, and the actual workings of an increasingly de-globalised economy. Between the countries that can adapt to the new age of insularity, and those that can't. And between the creaking nation state, and the newly empowered cities – whose influence has been turbocharged by new technology.

The world is not flat. It is very, very spiky.[3]

The classic rationale for globalisation rests on the work of the economists Adam Smith and David Ricardo, whose theories both make the case for freer trade. In Smith's argument, economic benefits arise from specialisation. If, instead of doing everything, we focus on doing one thing, we'll be more productive. That extra productivity can be used to trade with another specialist. I grind the wheat, you do the baking; between us, we end up with a loaf of bread. In short, trading has generally been more productive than an emphasis on self-sufficiency.

Writing in the early nineteenth century, David Ricardo developed Smith's ideas to emphasise the importance of 'comparative advantage'. In Ricardo's view, nations should export whatever they are relatively good at producing. If a country has extensive coal reserves, it should focus on producing and exporting coal; if a country is rich in arable crops, it should focus on producing and exporting food. Add to that the different economic opportunities present in different societies – poorer countries will be better at low-skilled work like assembly; richer ones at design and innovation – and you have a useful combination: by exchanging with each other, all countries become better off.

Those of us born in the late twentieth century inhabit a world of global trade that Smith and Ricardo could only have dreamed of. The supply chains of this global world touch every part of our lives. A phone is assembled using lithium from Chile, aluminium from China and palladium from South Africa. A single aircraft wing might cross a border half a dozen times, shunted from one specialist factory to the next.

Consider your food. For many years, vegetables, fruits, fish, meat and dairy products have criss-crossed the globe in refrigerated containers. If you place an order with your online supermarket for tomatoes to be delivered at the end of the week, it is possible they are still in the ground in another country – likely Spain, in my case – when you click 'Add to basket'.[4] This is a result of the proclivities of the Spanish climate; we could grow tomatoes in Britain, but in Spain's sunny climes, you can grow many more with much less effort. And thanks to the miracle of global trade, a logistics company will collect them from a warehouse in Spain and carry them to your local supermarket's distribution centre. In Britain, the national distribution hub is known as the Golden Triangle, and it sits in the middle of the country. Trucks departing it can reach 90 per cent of the population in four hours. Orders are then sorted and distributed to the individual who made the order – usually about three days later.[5]

As we progress even further into the Exponential Age, the tendency towards global trade in physical products is being inverted. There would be no need to exchange all those goods if you could source everything that you need locally. In the case of food, we wouldn't go to all that effort if it was straightforward to grow food – whether tomatoes or bananas or pineapples – here in rainy Britain. And the new technologies of the Exponential Age create that very possibility.

High-tech entrepreneurs have started to bring farming closer to where the food will be eaten. Urban vertical farms, popular in Japan and spreading elsewhere, are unusually efficient. In this set-up, the traditional field is chopped up and assembled in stacks. A modern vertical farm may run to 12 or 13 storeys high, each with a floor area of a few dozen square metres. This method increases the productivity of each square metre of 'farmland': when built vertically, 40 metres of growing area can concertina to nearly 10 times that. Using AI systems to control lighting, water and heat drives even more efficiencies. Computer-controlled intensive farms do not require pesticides, or other chemicals.[6] Some require 80 times less water than traditional farms. Soil is eliminated in favour of hydroponics (where the roots dangle in water) or aeroponics (where a nutrient-dense solution is misted onto the roots). Rather than using ordinary greenhouse lights, with their wide spectrum of colours, some vertical farms shine only the precise wavelengths to which the vegetables respond. Not even a photon of light is wasted.[7] By using renewable energy (often supplied via solar panels on the roof of the building), their energy costs decline and their carbon footprint drops even further. Provided you have the resources to invest in the technology, these farms can be built more or less anywhere – Spain, Britain or beyond.

Historically, food needed to be transported from rural farms to urban centres. But the new technology of urban farming means this need not be the case. With their smaller footprints, farms can be closer to the mouths they feed – sometimes even in the city they serve. Montreal's 160,000-square-foot Lufa Farms greenhouse, the world's largest, sits directly on top of a distribution warehouse.[8] A tennis court is less than 3,000 square feet; Lufa would easily fit 50 of those. The proximity of Lufa to its consumers allows for fresher product, cultivated for nutrition. And many urban farms are following this template: built close to the retailer, so that the tomato practically rolls from its vine into your shopping bag.

As of 2020, vertical farms have a tiny share of the food market. But the market for high-intensity vertical farms is growing at more than 20 per cent per annum, on the march up our exponential curve.[9] The effects of this shift could be staggering. If, in the twentieth century, that ancient human problem – that you can only eat what is nearby – was solved by globalised logistics, then the twenty-first century offers an alternative solution. Today, you can use technology to transform what is actually nearby.

But there's another solution that is even more radical. New technology reduces our dependence on certain classes of commodities altogether. Let's turn now from kale to coal. For a hundred years, we have moved around vast quantities of fossil fuels to meet our energy needs. Cargo ships laden with coal, then tankers with oil, and finally refrigerated super tankers for natural gas – all move prehistoric energy from its source to giant power stations. Apart from the handful of nations with energy self-sufficiency, fossil fuels drive a large portion of world trade. They are so essential that the United States has kept an almost permanent military presence in the Persian Gulf to ensure the flow of crude oil continues unabated.

But renewables have now put every nation on a path to energy independence. Once wind turbines are installed or a solar farm is deployed, they require few raw materials – and, as we saw in Chapter 2, such power supplies are fast becoming ubiquitous. This shift to renewable energy drastically reduces the amount of 'stuff' that needs to be carted around. In 1998, the UK consumed 63 million tons of coal, three-quarters of which went into electricity generation and a third of which was imported. A mere 21 years later, coal demand for electricity had reduced by 94 per cent and imports were down by 70 per cent. This is combined with a wider trend, in which we get more out of the electricity we use. Between 1999 and 2019, British GDP increased 75 per cent – yet the amount of electricity the economy uses has declined by 15 per cent. We literally create twice as much wealth for every kilowatt-hour of electrical energy we use. And this is only one example – dozens of countries, from Germany to Uzbekistan and the Ukraine to the United States, have had similar experiences.

The shift away from fossil fuels and towards renewables reduces global dependence on fossil-rich nations. Solar energy, thankfully, is much more equitably distributed. While not every nation is rich in fossil fuels, solar energy is possible everywhere. The most solar-rich nation, Azerbaijan, only gets four times more sunlight per square mile of land than the most impoverished, Norway. That may sound significant, but it is a relatively minor variance. The equivalent density between the haves and have-nots for oil is more than a million to one.[10]

This shift is being charged not only by new forms of electricity, but by new methods of energy storage. In an age of green energy, storage systems become more important – after dusk, solar farms become useless, and so you need a way to store large amounts of electricity. In part, that's down to innovations like Energy Vault, the company behind the giant insect-like cranes we met earlier. But many of the new methods of storage bring electricity closer to home. Our electric vehicles can hoard electricity which could also power our homes and offices through so-called vehicle-to-grid systems. The average electric car stores about 50 kilowatt-hours of electricity: enough to run the typical British or American home for five days. It will become commonplace for our electric cars to lend their stored electricity to our homes when it is dark. Britain alone is forecast to have as many as 11 million such cars by 2030. If each owner were willing to share a bit of the surplus energy stored in their cars with their neighbours, it might cover the whole country's needs.[11]

Simon Daniel is an inventor whose work reveals the power of these newly localised storage systems. His first success was a folding keyboard he designed in the 1990s, just as the PalmPilot, an early pocket-sized tablet computer, was taking off. His latest adventure is to string together thousands of batteries to make a gigantic virtual power plant. For Moixa, his company, to buy the batteries itself, it would need large amounts of capital, perhaps running into the tens of millions of dollars. Instead, he's persuading owners of electric cars to connect to his network. Together, these idle car batteries form a giant virtual power plant. Daniel's platform manages them and uses sophisticated algorithms to balance usage across the whole network. At last count, he had managed to combine 20,000 batteries together in several Japanese cities.[12] That's enough to power 25,000 Japanese homes for a day. It is like alchemy – replacing a massive power station, smokestacks rising into the sky, with a web of cars, parked on driveways, keeping homes running as we sleep.

Economies don't just use oil for energy, of course. Derivatives of crude oil are used to produce pharmaceuticals and plastics, essential parts of everyday life. But this too could be about to change. Josh Hoffman is the CEO of Zymergen, the biotech start-up. His company combines advanced machine learning with clever genetics to persuade microbes to efficiently produce industrial materials. Their first product, Hyaline, will be used in the screens of smartphones. Hoffman's bugs use natural processes to elegantly grow the screen film, requiring much less energy than making plastics from oil. It is a method that effectively removes hydrocarbons from the production of prosthetic materials.

There's a caveat to this process, of course. While we are becoming less dependent on fossil fuels, the transition to a fully green economy might not be complete until at least the late twenty-first century. And at the same time, we may become more dependent than ever on a slew of new commodities: the lithium in our batteries, for example, and the rare earth metals that are needed to make precision electronics. Yet the wider trend is towards an economy based on ever less stuff. These trends – the re-localisation of commodity production, plus our decreasing dependence on some commodities altogether – mark a radical shift. Soon, we may be able to fulfil many of our material needs without relying as heavily on international trade.

Perhaps even more revolutionary than our newfound approach to raw goods is the shifting world of manufacturing. In the Exponential Age, manufacturing is becoming less about putting trainers, phones, car components or prosthetics onto standardised 20-foot containers and shipping them around the world. Instead, manufacturing is taking the shape outlined by Angelo Yu. The idea is shipped across the globe, but the building process takes place at a printer or fabricator close to the point of consumption. This could make much of the global network of factories, logistical supply chains and offices redundant. They become liabilities.[13]

This increasingly localised world of manufacturing is driven by the new norms of the exponential economy. For one thing, we are witnessing the surging importance of intangible assets, which we explored in Chapter 2. For many complex products, from computers to advanced pharmaceuticals, from phones to automotive components, we are largely shipping around ideas. A $1,000 iPhone contains less than $400 in parts. The remaining three-fifths of its original sticker price are the intangibles – the design, the orchestration and the brand.[14] As new manufacturing methods mature, that ratio will only move in favour of intangibles. The rise of new exponential technologies, including AI, will accentuate this shift in value. Cutting-edge manufacturing processes require fewer workers than older methods – which means, for the first time in decades, it makes economic sense to manufacture in places with high labour costs.

Take running shoes. For years, branded sneakers have epitomised all that globalisation has to offer. Designed in studios in Portland or Berlin, and promoted by global sports stars on social media, the shoes themselves are largely manufactured in the factory nations of Vietnam,

Bangladesh or Thailand. But this is starting to change. In 2016, the German sneaker maker, Adidas, opened its first small-scale factory in its home country. At 50,000 square feet, the Speedfactory in Ansbach, Bavaria, could produce half a million pairs of shoes a year – relying not on cheap labour but on robots, automation and 3D printing. This approach might even end up with a higher standard of product. Such a factory could produce more complex, personalised shoes for its customers.[15]

We are not yet living in this exponential future, however. Adidas were a little too far ahead of the curve – within three years, the company had wound the Speedfactory down and moved its technology back out to factories in Asia. There were some things the machines couldn't yet do that long-established suppliers could.[16] End-to-end additive manufacturing remains relatively expensive compared to traditional manufacturing – robots aren't yet cheap enough for these methods to always work.[17] But this bundle of high-tech manufacturing technologies is riding an exponential wave, of the kind we have seen many times already. 3D printing technologies are expensive because it is currently a low-volume game. But it will march down the curve of Wright's Law – getting rapidly cheaper, and in turn driving lower costs for complementary products such as the software needed to control these printing machines. The main 3D printing technologies are already improving at around 30 per cent per annum.

The potential of this new technology was demonstrated during the Covid-19 pandemic. As the virus spread, ill-prepared nations like the United Kingdom found themselves sorely short of masks and visors for their doctors and nurses. Global production facilities in China and Turkey were themselves under lockdown. Volunteer efforts, often led by schools and universities, turned their 3D printers to work. A citizen supply chain got to work producing kit.[18] My children's schools produced several hundred masks using their 3D printers. The masks were then whisked away to the nearby hospital, a mere 20-minute stroll away.

This trend is in its infancy. The global market for 3D printing is tiny, only just nudging $10 billion per annum in 2019. Yet it is growing fast. And it is finding uses among demanding customers: those making components for cars and planes, for instance. One analyst estimates that using additive manufacturing to make a small strut for an airplane might save $700 million in fuel costs over five years. Printed component parts use less material than traditional approaches – they weigh less and waste less.[19] If you are lucky enough to drive BMW's top-of-the-range electric sports car, the i8, you'll use a 3D printed part every time you open the roof. One of the key widgets that allows the roof to zig-zag its way open is a a 3D printed bracket, and the company has committed to producing tens of thousands of components using this new approach.

And so here we can make out a new system of global trade. Gone will be the world of poor countries manufacturing goods for rich countries, and shipping these products across the world. Instead, each rich country will begin to make its own goods at home, for a domestic market.

But as ever in the Exponential Age, the rewards of new technology are not evenly distributed. The diminishing dependence of rich countries on poor countries' commodities may fundamentally destabilise the economies of much of the developing world. And depriving poor countries of manufacturing income – as high-tech local manufacturing becomes the norm in Europe and America – could be even more ruinous. A chasm will emerge between the rich world, which has harnessed the power of exponential technology to meet its own needs, and the poor world, which has neither the capital nor the high-skill labour force to keep up.

Let's look first at commodities. As David Ricardo predicted, much of the developing world has played to its strengths by primarily exporting raw materials – to be developed into more complex products elsewhere. According to one recent history of the commodities trade, 'Most African countries export commodities, and little else.'[20] Africa's economic fortunes have consistently risen and fallen in line with commodities prices – in the 1980s and 1990s, low commodity prices wreaked economic havoc in many sub-Saharan countries; through the 2000s, high growth rates in many

African countries could partly be linked back to soaring commodity prices, themselves driven by growing Chinese demand for raw materials. Between 2001 and 2011, the economy of sub-Saharan Africa grew fourfold.[21]

All this poses a problem. If the defining force of the modern global economy is to be the re-localisation of production, how will the countries of the developing world get by? The economies of many such countries depend on high demand for imported raw materials in the rich world – if this plummets, such countries will be in a precipitous position.

Consider the consequences of the collapse in oil prices during the 2020 Covid-19 recession. As plane flights were cancelled and a locked-down global population stopped refilling their cars, oil prices went into freefall. Such a shift destabilised huge parts of the global order. As Atif Kubursi, an economics professor at McMaster University, put it, 'Saudi Arabia needs an $80-per-barrel price to balance its budget, realize its plans to diversify its economy and sustain a heavily subsidized economy. In the balance is the stability of the Saudi Arabian political system and current regime.'[22] Such moments herald a concerning future for international trade. Decreasing reliance on commodities could immiserate large areas of the global economy, bringing with it untold political instability.

This emergent gap between the rich and poor worlds will be exacerbated by the shift to localised manufacturing. 3D printing will put downward pressure on the value of world trade. Analysts at the Dutch bank ING reckon it might eliminate up to 40 per cent of world imports by 2040.[23] This sum is colossal – an estimated $22 trillion – and will wipe out a decade's growth in the trade of goods and services. And if such proximate, distributed manufacturing did take hold, it would transform the whole nature of the global supply chain. UPS, the pre-eminent global logistics company, has already started to invest in technologies that will enable it to print parts that its customers need in a trice – instead of delivering packages across the world.[24]

Thanks to the compounding effect of these technologies, finished products won't need to be shipped from Mexico or Bangladesh; they can be printed in a nearby facility. The result may be devastating for already-struggling parts of the developing world, whose economies are built on global demand for cheap labour. Manufacturing goods for the developed world has brought more than just jobs. Whole cultures, economies and societies depend on the continued demand for cheap trainers in rich countries.

Between them, the trends of localised manufacturing and local production of food and energy could erode the authority of the states and multilateral institutions that maintain the global order of trade. In the case of the US, the sheer cost of maintaining global security and stability has run to trillions of dollars over the past decades, and has meant an almost permanent state of war.[25] Since 2019, when the US achieved energy sufficiency through its investments in fracking – a new method of extracting natural gas from bedrock – the country has been more willing to see chaos in petrostates the world over. In the words of Nick Butler of King's College London, American energy security 'removes one central argument for intervention in areas such as the Middle East, reinforcing the view that the US has nothing to gain from sending its troops to fight other people's wars'.[26]

The effects of exponential technology will be even more transformative. The institutions of a globalised world require nations to keep talking to each other, trading and cooperating. When the better-off world disengages from the economic wellbeing of the poor, it makes the path of economic development less clear. And it's a familiar pattern in the Exponential Age: high-tech, rich economies thrive; others get left behind.

At some point in the latter half of 2007, a man or woman strode into the future. As this individual stepped out of a crowded bus, perhaps packed elbow-to-armpit with a dozen others, they marked the beginning of a new era: one in which more than 50 per cent of our species resided in metropolitan areas.[27]

Throughout human history, the majority of people lived in rural rather than urban communities. But from that day in 2007 onwards, more than half of us have been city-dwellers. This moment was a long time coming. Two millennia ago, Rome exceeded 1 million inhabitants, as did Alexandria. But cities did not subsequently achieve that scale again until London reached it in the nineteenth century. Our first 10-million-person 'megacity', New York, was born in the 1930s on the back of various technological enhancements – electricity, mass transit, modern sanitation, steel-framed skyscrapers, the safety elevator. It also depended on new supply chains that could provide the food and goods needed to sustain a population of that size.[28]

And yet we have consistently underestimated cities. They have largely been eclipsed in political terms – with power being held at a national level. Cities are neglected in the metrics and institutions we use to make sense of the world – gross national product, national populations, national languages, national anthems.

The emphasis we place on countries erases the fact that cities have long been the engine of wealth creation, scientific discovery, trade and culture. In the Exponential Age, this trend will not just continue but accelerate. At first glance, exponential technologies might seem to erode cities' significance. They may allow for new forms of remote working. Such remote work, the theory goes, could give a greater online voice to people based away from urban areas. But this is only one side of the story. As we progress through the Exponential Age, cities will become more important, not less.

The key cause is, once again, the rise of the intangible economy – and the effect it has on labour markets. Value in this intangible economy is created through highly complex products. Today, highly skilled knowledge workers are more in demand than ever. And these types of workers have never been evenly spread across a country. They cluster around physical institutions, namely universities, and the labs and resources they provide. And around those universities grow companies that employ the most-skilled individuals.

Once a hub for any specialism is established – be it software design, structural engineering or contemporary art – others with an interest will flock there. Complex technologies – from biotech to artificial intelligence to chip design – need not just one but many specialists working together.[29] And only cities can bring them together.

Most of the high-tech urban areas in the Exponential Age are living proof of the way talented people from across sectors flock to cities – a process known as 'agglomeration'. Every year, thousands of tech workers migrate from around the world to San Francisco. As a result, technology companies are willing to shell out for the eye-watering cost of doing business in San Francisco (or London, or Paris) because they are getting the benefit of an incredibly skilled labour pool. Otherwise, they would move to smaller towns or the countryside and enjoy a lower cost of living.

And San Francisco is just one example. Today's intangible-powered cities are more specialised than ever. Mumbai has Bollywood, Tel Aviv has cyber security, and Hsinchu, a mid-sized city in Taiwan, has become the global powerhouse for chip making. The Hsinchu Science Park, which covers 14 square kilometres (about five square miles) – about a quarter of the size of Manhattan, and less than one-thousandth the size of London – generated more than $40 billion in 2020.[30]

This accumulation of talent in urban areas doesn't just help existing businesses and industries, however. It leads to the emergence of wholly new ideas. The close proximity of thousands of people leads to innovation. As strangers mix, new ideas pollinate. Cities are serendipity in action. Before the rise of smartphones and Amazon, most people would venture to a library or a bookshop to grab a book. In the course of locating the volume in question, you would pass by thousands of different works on hundreds of subjects. Perhaps your eye would catch on something completely random but incredibly useful while you scanned the shelves. Life in cities is similar. As someone leaves their home for their daily commute to the office, they encounter millions of different variables. It's as if they are scanning millions of books while looking for the right one. Strangers are the critical ingredient here: as the eminent American urbanist Jane Jacobs put it, '[Cities] differ from towns and suburbs in basic ways, and one of these is that cities are, by definition, full of strangers.'[31] Consider the power of this mingling in a city like Hsinchu, home to many of the world's leading chip experts.

There is an exponential effect at play: once a city has been established as the go-to place for, say, AI research, the world's best AI researchers flock there. More great ideas, innovations and businesses develop. And so the city only becomes more powerful. In this way, cities take on a kind of perpetual motion – the bigger they become, the better they are at their specialities; the better they are, the bigger they become. Geoffrey West, the complexity scientist introduced in Chapter 4, characterises cities as one big positive feedback loop. As cities grow, the professional opportunities grow, the social life gets better, and the more attractive it becomes. This, in turn, leads to those same people putting their money back into the city. 'The bigger the city the more each person earns, creates and innovates and interacts – and the more each person experiences crime, disease, entertainment, and opportunity – and all of this is at a cost that requires less infrastructure and energy for each of them,' West says. 'This is the genius of the city.'[32]

But what, in practice, does this mean for the future of cities? The short answer is that they will continue to grow. This is particularly true in the developing world, where rates of urbanisation are lower than in richer countries. As we move further into the twenty-first century, we will see the growth of ever-more megacities of more than 10 million people – in Asia, South America and sub-Saharan Africa. By 2030, nearly 9 per cent of the world will live in just 41 cities.[33]

Take the Greater Bay Area, a megalopolis – a cluster of megacities – which includes Shenzhen, Hong Kong and Guangzhou, and whose population exceeds 70 million people. Its economy will increasingly depend on high-tech and emerging sectors, and the benefits large agglomerations bring for economic development. That same logic holds true in much of Africa: by 2100, Lagos will be home to more than 88 million people, Dar es Salaam to more than 70 million and Khartoum to almost 60 million.[34] Even a global pandemic seemed unable to stop the rise of urban life. The early stages of the Covid-19 pandemic saw a flight from the cities – *The Economist* reported that 17 per cent of Parisians left the French capital as the country went into lockdown in March 2020.[35] Yet cities bounced back with remarkable rapidity. Researchers at the Wharton School of Business found that most of the major shifts away from cities early in the pandemic were temporary. While some particularly expensive areas such as Manhattan might see a more long-lasting exodus, cities generally held on to their residents.[36]

As cities grow, in both size and wealth, it creates the prospect of a gap between urban citizens and the rest of the population. City inhabitants tend to be more highly educated and better remunerated than a nation's median citizen. In India, for example, urban dwellers enjoy more than double the average annual income of rural inhabitants.[37] In 2018, when measured across 24 European countries, the earnings gap between urban and rural was 45 per cent.[38] All this can lead to a gulf between urban citizens, the great contributors to and beneficiaries of exponential technology, and the rest of a country.

This tension often leads to a conflict between national and city-level governments: with each claiming the right to determine how cities are run. For many years, London was one of Uber's five biggest markets. But the company grew there with scant regard for the city's transport regulations. Having fallen short of some requirements on safety, the city suspended the firm's licences. Rather than support that decision, Theresa May, the prime minister (and national head of government) criticised the local authorities. Though she did not, in this case, have the power to overrule them.[39] This is a microcosm of a much broader tension between often liberal, rich, high-tech cities, and often less liberal, poorer, less technologically advanced national governments. For example, the *Financial Times* recently pointed towards a tendency for national and city executives to clash over immigration – as national governments have turned against migrants, cities in need of labour have taken a more open approach. And that makes them more welcoming to immigrants than their wider countries. As Samer Saliba, an adviser for the refugee NGO the International Rescue Committee put it, 'Nations talk and cities act.'[40]

Of course, it would be foolhardy to extrapolate too much from these early clashes between urban and national. But we can begin to make out the contours of an Exponential Age society in which there are constant conflicts between cities and nations. Exponential technologies allow cities more autonomy and self-sufficiency in electricity, trade and food. The exponential economy favours the complex, high-skill activities that are best supported by large, diverse urban populations. If the industrial age of nineteenth and twentieth centuries cemented the importance of the nation state, the Exponential Age is shifting much of that significance to great cities. The result is a tension over where decisions about many aspects of the daily lives of more than half of humanity should occur: at the level of the nation state, or closer to home, in the city.

John Perry Barlow, the former lyricist of the Grateful Dead, was also an internet rights activist for the Electronic Frontier Foundation. At the World Economic Forum's Davos summit in 1996 he issued a Declaration of the Independence of Cyberspace:

> *Governments of the Industrial World, you weary giants of flesh and steel, I come from Cyberspace, the new home of Mind. On behalf of the future, I ask you of the past to leave us alone. You are not welcome among us. You have no sovereignty where we gather.*
>
> *We have no elected government, nor are we likely to have one, so I address you with no greater authority than that with which liberty itself always speaks. I declare the global social space we are building to be naturally independent of the tyrannies you seek to impose on us. You have no moral right to rule us nor do you possess any methods of enforcement we have true reason to fear.*[41]

Barlow was not alone in thinking that the internet transcends national boundaries. It represents one of the most important symbols of globalisation. The web passes across borders like a pulmonary system carrying the oxygen of globalisation: not merely facilitating the trade of goods but also the percolation of ideas. It is aloof, above the worldly concerns of border guards, passport agents and customs checks.

And in Barlow's time, the internet did indeed seem to subvert the logic of the nation state. It undermined the most basic building block of the geopolitical order: territorial sovereignty, the notion that a nation controls what goes on in its borders. Behind the Iron Curtain and in religious autocracies, the early internet was a place to explore ideas that couldn't be expressed in the national press. The elimination of the government-controlled middleman, the radio or TV broadcaster, meant protesters could express themselves freely. The failed coup of August 1991 that led to the dissolution of the Soviet Union is a great example. Protestors were able to signal across the fledgling internet that there was a coup underway, even as the unsuccessful rebels shut down CNN and other mass media.[42] More broadly, the constant criss-crossing of data – in which ideas might spread from Edinburgh to Évian, Mumbai to Manhattan – helped build a uniquely international culture online.

This isn't to say that the internet was without national markers. The norms of the early internet had a distinctly American hue. The key protocols, standard procedures, and institutions of governance were almost entirely created by Americans, often sponsored by the US government. And that brought with it a particular set of preoccupations: academic, increasingly liberal, and sceptical of centralised authority. In general, however, the internet promised a global, borderless future.

A quarter of a century after Barlow's speech, that ideal is in tatters. The internet is turning into a 'splinternet', fragmented into regional and national spheres. During the 2010s, governments started to find new ways to bring the internet under the thumb of nation states. Many authorities feared that their citizens' data might be turned against their people or the state if allowed to flow beyond their borders. The first to understand this were authoritarian regimes. For those living in China, internet use is restricted by vast armies of censors – and by the 'Great Firewall', which blocks access to large chunks of the global network. Citizens in Iran, Syria, Turkey, Egypt, the Philippines and many other countries regularly have their internet access surveilled or blocked.[43]

More recently, this dynamic has spread to liberal democracies. In many cases, their reasoning is understandable. European data protection laws, namely the General Data Protection Regulation – which imposes a strict set of rules on how companies collect and use data – is perhaps the best example: a benign attempt to protect citizens' digital rights, which nonetheless places hard limits on how easily data can spread across the world. Elsewhere, governments' actions have a more authoritarian tinge. India's national e-commerce policy, for example, argues that 'the data of a country is best thought of as a . . . national asset, that the government holds in trust' – a statement that Barlow would likely have been appalled by. The Indian government looks set to create restrictions on how data might be transferred out of the country, leading many technology companies to physically build 'data farms' – that is, giant warehouses of servers – in the country.[44] This trend looks set to continue. In 2020, several foreign-owned applications, such as TikTok, were banned in India.

As a result, the internet looks increasingly localised, and far from the 'global social space' of Barlow's era. Data protection laws are complicated, inconsistent and commonplace. One London law firm now summarises such rules in a handbook, running from Angola to Zimbabwe. It weighs in at a thumping 820 pages.[45] In the decade leading up to 2016, the number of countries that introduced territorial rules around data nearly tripled, to 84.[46]

The story of the de-localisation and re-localisation of the internet is perhaps the neatest summary yet of the trajectory of the Exponential Age. Digital technologies have the potential to transcend national borders; but, as we have seen, they are just as likely to strengthen them. And that is especially true at a time when many global governments are turning back to nationalist politics.

All this poses a problem for policymakers. The institutions that guide global policy and security were built during an era in which it was assumed that locality was not only dead, but perhaps didn't matter. From geopolitics to global economics, institutions like the IMF and the WTO have long been the missionaries of globalisation, establishing a template for what participation in the world economy would like. These assumptions seem out of place in a world of national self-sufficiency in energy and commodities, localised manufacturing, and cities of ever-growing importance.

As a result, we need creative responses to the growing divisions in the new world. Those between high-tech, rich and increasingly independent nations, and their poorer, less technologically advanced counterparts. And between growing liberal cities powered by Exponential Age technologies, and more conservative, less cosmopolitan national polities.

The first gap may be closed by an acknowledgement of the continued importance of international cooperation, for developed and developing countries alike. There remain challenges that can only be handled at a global scale, whether that be the threat of anthropogenic climate change, future pandemics or – as we will discover in the next chapter – cyber security. But if the basis for international teamwork isn't trade, what can bring about that cooperation in practice?

One answer involves establishing what institutions are best equipped to tackle emerging global issues. We might create bodies that foster cooperation, to complement the global trade organisations of old. One such might be a World Data Organization – an idea advocated by the political scientist Ian Bremmer. Such organisations would need to adopt a different approach than the globalising institutions of yore like the IMF. A global data body could coordinate a consistent approach towards artificial intelligence, citizens' data and intellectual property. It could keep data – needed for healthcare, industry, climate change and research – flowing between countries, even as digital walls go up.[47] This might, at first, seem like an inadequate response to forces as significant as the re-localisation of supply chains. But keeping data flowing between borders – in the form of radical new inventions, or clever software patches – would help prevent economies falling into siloes. Even if production processes become local, that doesn't mean the ideas underpinning production must too.

A similar goal – preventing re-localised economies resulting in new barriers between nations – might be achieved through 'digital minilateralism'. That means the cooperation of small groups of nations on questions of how to regulate the digital world. One pioneering example of this is the Digital Nations group, founded in 2014 by a grab bag of countries, none of which even shared a border: Estonia, South Korea, Israel, New Zealand and the United Kingdom came together to cooperate on common digital projects, including developing principles of how artificial intelligence might be used by governments. Minilateral approaches benefit from being free of the sclerotic bureaucracy of universal organisations like the UN. Drawing as they do on a smaller pool of countries, minilateral organisations may be more agile and thus more able to make meaningful progress, according to Cambridge University researchers.[48] For such an approach to close the exponential gap between rich countries and poor, it would need to invite many more players from the developing world to participate. But it does demonstrate that re-localised economies need not lead to isolationist, uncooperative nation states.

The second growing divide – the one between nation states and cities – needs similarly ambitious solutions. The growing economic, cultural and demographic power of cities requires recognition. Cities are often at the forefront of tackling problems caused by exponential technology. It was places like London that first had to contend with the sudden growth of gig-working platforms such as Uber. Barcelona was among the earliest to reckon with the explosion of digital accommodation marketplaces like Airbnb, and all the economic effects that brought. And, as we have seen, cities will be the engine of the Exponential Age economy.

The solution may be to develop more federal models of national politics, which give more power to regions and cities to manage their own affairs. They need the ability to attract people and regulate the quality of their citizens' lives, by increasingly governing their own energy, resources and climate agendas. This might seem likely to exacerbate the rural-urban divide, but in fact it may help close it. Many of the divisions in our society result from cities and nations – which have vastly different economic interests and political outlooks – being tied together under a single, over-centralised government. Good examples of an effective, federalised approach to urban government are hard to find. However, many cities seem up for the challenge. Many municipal governments have recently started to work together to identify the shared policies they need. The C40 initiative brings mayors together to discuss climate change; the Mayors for a Guaranteed Income is a coalition of American cities advocating for universal basic income.

These forms of revitalised urban governance are deeply necessary. Exponential technologies take our apparently flattened two-dimensional world and, like a pop-up map, make valleys and peaks suddenly visible. And this is a terrain that our current institutional arrangements, from our approach to trade to how we think about local governance, are ill-equipped to deal with. We may find we have little choice but to accept the re-localising effects of new technologies; what we can do is mitigate the risk of this process leading us into an era of political and economic disorder.

But if we do find ourselves entering a newly disordered world, what will that mean for global conflict? As countries become increasingly economically independent, and tensions mount between urban and rural areas, it seems plausible that wars between and within countries will become more common. And, for better or worse, our approach to warfare is being disrupted just as dramatically as is our approach to geography.

7

The New World Disorder

Toomas Hendrik Ilves had just returned from Moscow when the trouble began. The president of Estonia, Ilves had been visiting Russia to attend the funeral of the country's former leader Boris Yeltsin. After several days' travel and official duties, he might have hoped for some respite. But no sooner had he returned to his official residence in Tallinn than Estonia was plunged into crisis.

It was April 2007, and through the year tensions between Estonia's Russian minority and the authorities had been growing. Ilves's recently elected government had taken steps to relocate a Soviet-era statue away from the centre of Tallinn. The young nation, independent from 1991 after six decades of Soviet rule, was seeking to establish its cultural sovereignty. Yet to many Russians in Estonia, the decision to move the statue represented an assault on their heritage. Now riots had broken out in the capital. The president's security detail feared for his safety, and he was whisked away to a farm outside the city as a precaution.

That was just the beginning. The next morning, deep in the countryside, Ilves found himself unable to log in to the government's internet systems. The local newspaper website was offline too. His online bank was unresponsive. Eventually, Ilves tried some American websites, CNN and the *New York Times*. They loaded seamlessly. It was then that he realised what was going on.

Estonia had been hit by one of the most successful cyberattacks in history. These digital salvoes targeted the entire national infrastructure of the country. Shortly after the mob took to the streets, a wave of botnet and denial-of-service attacks targeted the media, banking and government services, taking them offline. Estonians were unable to withdraw cash from ATMs. Email services went down. Broadcasters were frozen out of their systems; trusted local news services fell silent. The attack was compounded by a coordinated wave of misinformation – pro-Russian hackers flooded news sites with fake news about the events as they were unfolding, adding an element of hysteria to an already fraught situation. The Estonian government, which led one of the most digitally connected countries in the world, was forced to pull the plug and disconnect the country from the global internet in order to end the attack.

To this day, it isn't quite clear who exactly was culpable for the attack. Many government officials blamed the Kremlin – the attacks came from Russian IP addresses, and online instructions that directed the attack were in Russian. However, in the Exponential Age it is rarely straightforward to find a single culprit: the trail of evidence is too complex. One Estonian government official later told the BBC the attack 'was orchestrated by the Kremlin, and malicious gangs then seized the opportunity to join in and do their own bit to attack Estonia'.[1]

Estonia is a member of NATO. An attack launched by tanks and aircraft would have triggered Article 5 of the NATO treaty, and members of the alliance would have flocked to Estonia's aid. But faced with this very modern kind of warfare, the right response wasn't so clear. The Estonian government considered invoking the article but lacked absolute proof that the attacks were from the Russian state. It was not clear whether Article 5 would, at that point, cover cyberattacks anyway.

But whoever was responsible, the attack changed the world. It demonstrated the power of this form of attack to governments globally. And it presaged what the future of conflict would look like.

As exponential technologies march forward, they bring with them two changes. Just as the price of computing has collapsed since the 1970s, so the cost of building certain types of weapons is declining precipitously. Computer viruses and other malicious codes are simply software – they are driven by the same price dynamics that give us better laptops and cheaper mobile phones. More nebulous forms of attack become more feasible too. In the twentieth century, for a state to orchestrate a wave of misinformation they would have had to infiltrate newspaper offices, radio and TV studios and broadcast transmitters. Today, it's merely a matter of making malicious posts go viral on social media. Even in the physical world the cost of warfare is diminishing too. The price of military drones has declined by a factor of a thousand in a decade or so. All these forms of attack are cheaper than their twentieth-century predecessors – and none puts the attacker in the immediate line of fire.

While attackers have gained many new, often cheap, forms of offensive, defenders find themselves much more vulnerable. When few of us owned a computer, few of us could be targeted by some kind of internet ruse or cyberattack. As digital devices have proliferated, so too have our vulnerabilities. Every smart device you buy opens a potential security loophole. A Wi-Fi-connected colour-changing light bulb may be fun, but it may also be an appealing device for an attacker to latch on to. Businesses in all types of industries, including the critical infrastructures of healthcare and energy, are connected to the internet – and that makes them accessible to bad actors. Combine this with the growing number of people online, connected via misinformation-prone social networks, and you have a huge number of targets.

In security terms, these targets are collectively known as the 'attack surface' – the total number of vulnerabilities that can be exploited by a hostile actor. A medieval king in a castle, protected by a moat with a single drawbridge and facing archers, swordsmen and primitive ballistics, had a very limited attack surface, given the capabilities of his attackers. But as time went on, the attack surface expanded – because the weapons that could be used against it increased in number. The arrival of trebuchet and then cannons rendered the thick walls meaningless, and the development of aircraft made the height of the ramparts redundant. No one today would think of using a castle to defend from a modern assault. This, in short, is what the technologies of the Exponential Age are doing to the modern state, not to mention modern businesses. Our security systems look increasingly feeble, our attack surface ever more vulnerable, when set against new military technologies.

All of this leads to a classic exponential gap. The speed of technological change has created new opportunities for our adversaries. But the channels we might use to attenuate conflict, or the resources we can draw on to reduce the severity of attacks, have not kept up. On the one hand, methods and sites of attack proliferate. On the other, you have national security infrastructure forged iteratively over centuries. As defending ourselves becomes more complex, even the best militaries, which are mostly moving at a sluggish pace, struggle to adapt.

And the gap is not just between new forms of warfare and our existing defences. It also encompasses the norms and laws governing war zones. Over the centuries, we have established rules of engagement on the battlefield and between nations – from the Geneva Convention to the Treaty on the Non-Proliferation of Nuclear Weapons. Digital warfare will require similar protocols. What are the rules governing an army of malicious hackers? Or the way a hoard of drones interacts with citizens? Away from the physical battlefield, can countries respond to a cyber-attack with a bombing raid? Would it be reasonable to respond to a wave of misinformation marshalled by an opponent with a wave of missiles?

All this uncertainty comes against a backdrop of growing political instability, both between and within countries. As we saw in the last chapter, the world is getting spikier. New technologies lend themselves to localised production and less global trade. And less trade might mean more conflict. Writing in 1848, the British liberal philosopher John Stuart Mill characterised 'the great extent and rapid increase of international trade' as 'the principal guarantee of the peace of the world'.[2] This has long been a controversial idea – the century that followed Mill's statement was defined by unprecedented levels of global trade and culminated in the two biggest conflagrations in human history. But, in general, the principle holds. When nations trade, they become economically interdependent. Conflict becomes an inconvenience – it would get in the way of business.

In one dizzyingly enormous study by the Asian Development Bank, researchers analysed hundreds of thousands of conflicts between 1950 and 2000 and concluded that 'an increase in bilateral trade interdependence and global trade openness significantly reduces the probability of military conflict between countries'.[3] Unsurprisingly, then, the turn against globalisation may well lead to greater conflict between nations. One 2010 estimate by a three-decade veteran of the Central Intelligence Agency suggested that deglobalisation might raise the risk of war for any given country in the subsequent 25 years by more than a factor of six.[4]

At the same time, conflicts within nations are becoming more commonplace. This dynamic long predates the moment in the 2010s when we tipped into the Exponential Age. Writing in the 1990s, the academic Mary Kaldor coined the term 'new wars' to describe conflicts after the fall of the Soviet Union. For most of the late twentieth century, wars had been fought between states and their proxies, probably over ideological disputes – namely that between communism and capitalism. The objective was to weaken an enemy state rather than to target its population. Since the 1990s, however, the nature of war has become more fragmented. New wars are led by non-state actors, who are often driven by religious and cultural identities as much as by ideology. Citizens are viewed as a legitimate target. And these conflicts are more likely to unfold within countries than between them. Think of the interethnic conflicts in 1990s Yugoslavia, or the rise of Islamic fundamentalist forces like ISIS.[5]

As we charge further into the Exponential Age, new wars have become even more common. By 2018, the number of non-state conflicts tracked by the Uppsala Conflict Data Program exceeded 70, more than double the number a decade earlier.[6] It would be excessive to say that exponential technologies have caused this shift. But Exponential Age trends have no doubt contributed. Growing urban-rural divides, the increasing weakness of intergovernmental bodies, and the declining cost of military technology have all drawn non-state actors into the fray.

As a result, the next few decades could be an era of grinding, constant conflict. Nations' ability to defend themselves will look increasingly flimsy, just as our economic and political order becomes more unstable. Sir Richard Barrons – former commander of the British Joint Forces Command, who undertook tours of duty in Northern Ireland, Kosovo, Iraq and Afghanistan – once explained the scale of the problem to me. 'Many of the things we set great store by, certainly in the arena of conflict,' he said, 'are being literally blown away by the way technology is changing how war is fought.'[7]

In 1981, the Israeli Air Force's Colonel Ze'ev Raz led a formation of eight F-16 fighters on a secretive mission to Iraq. On 7 June, his squadron of small manoeuvrable planes – designed mostly for dogfighting but on this occasion heavily laden with Vietnam-era gravity bombs – took off from Etzion Airbase on the Sinai Peninsula.[8] Before them lay a three-hour, 3,200-kilometre round trip which would reach the very limits of the planes' operational range.

As they flew over hostile airspace, they were forced to masquerade as Jordanian and Saudi planes to avoid local air defences.[9] Their target? The Osirak nuclear reactor – part of Saddam Hussein's programme to develop nuclear weapons. For all the risk borne by the airmen, the bigger risk was the potential of nuclear escalation in the region and, in the view of the pilots, the threat posed to the existence of their home nation.

Operation Opera was both expensive and risky. The aircraft, supported by escort fighters and refuelling tankers, each cost tens of thousands of dollars an hour to run. The planes were worth millions of dollars apiece. The whole operation could have gone disastrously wrong. And if it had gone even slightly akimbo, the Israelis may have had to contend with one of their airmen getting trapped deep in Iraqi territory.[10]

However, the mission proved an unqualified success. Raz and his team delivered 16 tons of explosives onto the hardened concrete shell protecting the reactor in under 90 seconds. It was obliterated and, with it, so were Hussein's nuclear weapons ambitions.

Fast-forward four decades and the Israeli state is still concerned about nuclear proliferation in nearby nations – but today their focus is no longer on Iraq. Since the turn of the millennium, Iran has harboured ambitions to develop nuclear weapons. And in the Exponential Age, Israel's responses have taken a different form.

The Natanz nuclear site is an industrial complex a few hundred miles south of Tehran, used to enrich uranium. Starting in early 2010, the site suffered a series of perplexing mishaps – around 1,000 centrifuges, a tenth of the site's contingent, were destroyed. But the perpetrator was not a squadron of Israel's feared top guns and their 20-ton aircraft. It was Stuxnet, a piece of malicious software code. The bug had buried into Iranian firms' networks for a few days before arriving in Natanz. Designed specifically to target the software running on electronics made by the German firm Siemens, Stuxnet spent a month gently tampering with the operation of Iran's computers. The goal was to damage the delicate centrifuges that were being used to purify uranium for military use.[11]

The effects went far beyond the digital world, destroying much of Iran's nuclear capabilities. And it all happened without a 'kinetic' attack – that is, lobbing a bomb at an enemy. While nobody has ever officially claimed responsibility for Stuxnet, it has been widely alleged to be a joint creation of the Israeli and US governments.[12]

All states have vulnerabilities similar to the Natanz nuclear site. Today, everything has – or will soon have – a digital interface, probably connected to the internet. One effect of this transformation is to increase the number of attack surfaces within governments – be they nuclear reactors or military databases. But the shift affects civilians too. Millions of civilian-targeted cyberattacks happen every day; and 96 per cent of businesses in Britain suffered a damaging cyberattack in 2019.[13] Data leaks are one of the most common consequences of cyberattacks – personal and customer data can be sold on the black market, or used to compromise more computer systems. In 2020, more than 37 billion data records were leaked worldwide, a 46-fold increase on five years previously. Three in five of such breaches, about six a day, occurred within the United States.[14] This is a change we are ill-equipped for – whereas most of us still worry about whether we remembered to lock the front door, few of us have the ability to take even basic measures to protect ourselves from cyberattacks.

For every new attack surface, there is a new attack method. Nicole Eagan, CEO of cyber security company Darktrace, told me about the case of an unsuspecting corporate exec who found the computer in his swanky electric car had been used to transmit information to a malfeasant. The car had been charging in the company car park; in the process, it connected to the company's Wi-Fi and sent data about its owner's driving habits to the manufacturer. This internet link was enough for the car's data to be hacked. Some days later, the attack was repeated on a different car in a different city.[15] In other cases, hackers have found ways of 'spoofing' global positioning systems, to convince ships' systems that they are thousands of miles from where they are – useful if you want to disrupt global flows of trade.[16]

And while those attacks on civilian networks seem relatively minor, others have been catastrophic. In 2017, a group of Russian agents – known collectively as Sandworm – launched an operation to debilitate Ukraine's banking system. They released a piece of malevolent software, or 'malware', into the servers of Linkos Group, a Ukrainian software company. Linkos regularly sent updates of its software over the internet to companies all over Ukraine. It was a perfect place to inject malware and hit the country's computer infrastructure.

In that, the attackers were successful. But thanks to the global connectedness of the internet, the effects were felt elsewhere too. Within a few hours, the malware 'NotPetya' had debilitated hospitals in the US and a factory in Tasmania. It would ultimately spread to PCs in 65 countries.[17] Some of the most badly hit victims were global companies: Maersk, the Danish shipping giant; Mondelez, the world's eleventh-largest food company; and RB, the household products titan. The disruption caused billions of dollars of damage – by some estimates as much as $10 billion.

This proliferation of digital attack targets and methods is fundamentally changing the face of contemporary warfare. It is bringing new and unexpected actors into international military conflicts. Take North Korea. For all of the bluster of Kim Jong-un's regime, the country's politics have historically been relatively insular – while the government is fond of its military parades, its involvement in international conflicts has generally been limited. It certainly hasn't been able to afford many physical operations beyond its own borders.

But that is changing fast. The North Korean cyber army included anywhere between 5,000 and 7,000 troops as of 2016 – all under the direct command of leader Kim Jong-un.[18] The importance of this organisation lies not only in its attempts to disrupt South Korean infrastructure, but in its involvement in illegal state activities. The North Korean cyber force has drawn blood with online scams, fraud and attacks on online banks and cryptocurrency exchanges.[19] International organised crime and cyber warfare are thought to have fuelled the growth in North Korea's GDP since 2015.[20] A UN report estimated that North Korean cyber experts have illegally 'raised' up to $2 billion for the country's weapons programs. This is a much cheaper form of extraterritorial activity than the traditional modes of warfare – which involve sending troops to the other side of the world. All the North Koreans need to wreak havoc is a computer and a talented hacker.

This dynamic means many relatively poor nations can have a disproportionately impactful foreign policy. In 2012, a particularly malevolent virus, Shamoon, struck oil and gas services across Qatar, the UAE and, in particular, Saudi Arabia. The hard disks of more than 30,000 computers were rendered inoperable.[21] The attack wiped out the drilling and production data of the Saudi state oil company Aramco, and hit its offices in Europe, the US and Asia. It took the company almost two weeks to restore its network.

Security researchers traced the virus's origin to APT33, a hacker group closely associated with the Iranian government. The country's use of cyberattackers is characteristic: the Iranian regime has long favoured foreign policy exploits that are cost-effective and deniable.[22] Iran's military spending is lower than its three main rivals, Israel, Turkey and Saudi

Arabia – but cyberspace gives them an opportunity to level up. Hackers give the Iranian government an edge, and it is one they are keen to exploit.

All this means we are entering a different world of conflict: one with radically new targets, forms of attack and hostile actors. But most of the ways we conceptualise warfare and defend ourselves remain distinctly twentieth-century. These new methods of offense don't map neatly on to our traditional conceptions of sovereignty and war and peace, developed over centuries. The Treaty of Westphalia of 1648 gave birth to the principle that the world was divided into sovereign nations with clear borders. The Geneva Conventions of 1929 and 1949 helped determine what was and was not a legitimate target for military conflict. None of these treaties are equipped to regulate a world in which thousands of actors seek to wreak havoc through digital subterfuge without any official acknowledgement of war.

Sandworm, for example, is a shadowy entity – allegedly part of Russia's Main Intelligence Directorate, better known as the GRU.[23] They were the group behind NotPetya – and are also likely behind Russia's meddling in American elections, the downing of a Malaysian plane flying from Amsterdam to Kuala Lumpur which killed 300 passengers,[24] and the use of a nerve agent in a failed assassination attempt in the UK in 2018. But while sometimes they have clear strategic objectives, at other points their only goal seems to be chaos. In the words of journalist Andy Greenberg, an expert on Sandworm, they are part of a 'reckless, callous military agency [acting] like cutthroat mercenaries'.[25] Sandworm had no real interest in Danish shipper Maersk or indeed any of the many Russian companies downed by NotPetya. It merely sought to create disorder in Ukraine. To do so, it exploited holes in the Microsoft operating system, a piece of software written by an American firm with a global footprint. So was this an act of war against Denmark or the US? It's hard to say.

The inadequacy of our rules-based order is compounded by the inadequacy of existing forms of national defence. Maersk, one of the shining jewels of Danish business, was crippled by Sandworm's cyberattack. All

49,000 of the company's laptops were destroyed in the attack, and more than 1,000 business applications were ruined – all in the seven minutes it took the virus to propagate inside the firm.[26] But the Danish state could do little to help the company; in the end it was the company's own IT staff, not the military or police, who stopped the attack. Likewise, when the virus WannaCry spread across the globe in 2017, it was two private security researchers who came to the rescue – teaming up with the American firm Cloudflare to disable the attack.[27]

These are not exceptional cases. Most nations' cyber defences remain feeble. The US military formed its Cyber Command in 2010, two years after classified information in the Department of Defense was compromised, and three years after the savage Russian cyberattack on Estonia. Yet it was not until 2018 that Washington adopted a principle of 'defending forward' – an approach that allowed its digital soldiers to mount offensive operations. Britain only launched its own National Cyber Force – an amalgam of its military, top spies, codebreakers and defence scientists – in 2020.[28] So far, these defensive capabilities have proven to be limited. In 2020, the American government realised that SolarWinds, a technology firm popular among government departments, had been the victim of a cyberattack that had gone undetected for months. Foreign hackers had been spying on private companies and departments like the US Treasury and the Department of Homeland Security. They were able to access tens of thousands of systems for months on end.[29]

Never have nations looked so incapable of defending their national interests. And never have outlaws, acting on their own or at the behest of a state, appeared so powerful.

Veles, in North Macedonia, seems like an unremarkable place. A city of fewer than 50,000 inhabitants, it lies nestled in the hills some 60 kilometres (37 miles) to the south-east of Skopje, the country's capital. And though it hosts a handful of beautiful churches and the odd tourist,

Veles has, for most of its history, done little to capture the imagination of the global public.

That all changed in 2016, when the press began to report on a booming new industry in Veles: fake news factories. In the run-up to the US presidential election of 2016, this tiny, decaying town became home to 140 websites focusing on American politics – most of them aggressively pro-Trump. Local teens were enlisted to conjure up or rewrite right-wing news stories, which were then published on social media. These stories were wildly popular, with some garnering hundreds of thousands of shares on Facebook. They were profitable too. One website owner was reportedly soon making upwards of $2,000 per day from online advertising.[30]

Those in Veles had little understanding of US politics. In many cases, they knew nothing about the topics they wrote about. But not only were they able to report on US politics; they were able present their analyses to millions of people.

The phenomenon of viral misinformation, sometimes created by state actors, is another manifestation of the growing number of attack surfaces in the Exponential Age. Of course, the use of information operations to motivate an army or dispirit the opposition is as old as the hills. Muhammad Saeed al-Sahhaf, the Iraqi diplomat, earned the moniker 'Comical Ali' for appearing on live TV in 2003 to claim victory for Saddam Hussein's forces – almost immediately before Baghdad fell to American troops. But in the Exponential Age, misinformation has become much easier to produce. And for this misinformation to achieve its aim, it doesn't need a skilled hacker or any complex code. It just needs members of the global public to believe it.

A decade before the 2016 US election, the idea of computational propaganda that could influence national politics seemed fanciful. Governments that managed to control the flow of information through their countries, like the Chinese, did so through gargantuan efforts, and only within their borders. The firm I founded, PeerIndex, analysed flows of content across Twitter and Facebook for several years between 2009 and its sale in 2014.

In those early years, we had to deal with very little coordinated misinformation. Automated social media accounts (known as 'bots') were common nuisances, usually programmed to amplify commercial scams for vitamins or erectile dysfunction treatments. But they posed little real threat. We built simple systems to detect and ignore them.

But by the time I had sold my company, the situation had started to change. By 2014, viral misinformation was a small but tangible problem. From there on in, the problem grew rapidly. According to a study from the Massachusetts Institute of Technology, the number of English-language tweets that made false claims on Twitter rose from almost negligible levels in 2012 to become a significant threat by 2016. In that period, the number of false claims increased by a factor of 30. Falsehoods, the study found, spread across the network six times faster than the truth.[31]

The technologies of the Exponential Age make the spread of misinformation much more likely. Anyone can create a website, and anyone can post content to it. It is incredibly cheap to create networks of automated social media accounts to amplify messages. Social networks are fertile ground: content flows freely, without the steady hand of an editor or fact checker. The platforms themselves reward content that disgusts, shocks or delights. And where online platforms use algorithms to recommend content, they do so in ways that often amplify pre-existing beliefs (a phenomenon we'll explore in the next chapter). In short, an unhealthy combination of exponential forces – new information networks, viral network effects and AI – combine to help lies spread.

And that means fake news is everywhere. In 2019, a survey of citizens in 25 countries found that more than 4 in 5 thought they had been exposed to fake news. Nearly 9 out of 10 of those had initially thought a fake news story was real (your author is in that majority).[32] Nearly half of Turks and more than 30 per cent of Koreans, Brazilians and Mexicans say they are exposed to fake news on a weekly basis.[33] During the 2020 US election, social networks became breeding grounds for malevolent actors fomenting conspiracies about vaccines or the origins of Covid-19 – bolstered by the increasingly erratic tweets of the then president Donald Trump.

Some of this fake news is created purely for profit – think of that $2,000-a-day website owner. But much is orchestrated by nation states seeking an advantage over an adversary – it is used as a form of attack. It is this characteristic that makes some misinformation not just a nuisance but a type of hostile military action. And this category of misinformation is increasingly common. In 2017, Oxford University researchers found evidence that 28 countries had undertaken some form of online disinformation operation. By 2019, this number had risen to 70.[34] Russia topped the league table, responsible for 72 per cent of all foreign disinformation operations between 2013 and 2019.[35] But the Chinese have taken a leaf out of the Kremlin's playbook, rolling out large-scale misinformation operations in the Asia-Pacific. In 2020, Facebook announced that it had uncovered a Chinese misinformation campaign in the Philippines, which seemed to have been designed to promote pro-China politicians.[36] Countries like Thailand, Iran, Saudi Arabia and Cuba have also been accused of creating networks of misinformation.[37] When used by states in this manner, misinformation becomes disinformation – that is, information that is actively malicious rather than merely inaccurate.

These disinformation campaigns can have a troubling real-world impact. In 2016, Russian internet trolls were able to organise a protest and counter-protest in Houston, Texas. More than 60 people showed up – not the instigators, though, who were sat at their computers thousands of miles away.[38] But the consequences can be much more harmful than the odd small protest. The World Health Organization described the 'infodemic' that was accompanying the Covid-19 pandemic in Iran, where 700 people may have died from methanol poisoning following online rumours encouraging them to drink the substance to rid themselves of the virus.[39] In the UK, dozens of telecoms engineers upgrading mobile phone networks have received death threats because of a cockamamie theory that 5G masts spread the virus.[40] And in the US, hostile state actors – particularly Russian ones – mounted concerted campaigns seeking to weaken America's response to the pandemic. Tactics emphasised 'weaponising' pre-existing divisions in the US.[41] The goal? To delegitimise efforts to control the spread of the disease and the subsequent vaccination programme.[42]

But if exponential technologies were one catalyst of the misinformation boom, another was the inability of states to effectively counter lies online. Governments have been startlingly late in taking responsibility for combating misinformation. Only in 2018 did the UK government begin to set up a national security unit to tackle disinformation, and even then they were criticised for the slowness with which they proposed to do so.[43] Meanwhile the US had, by the end of 2020, no coordinated counter-disinformation strategy steered by any branch of government.[44]

But if you think this means the future of conflict will be wholly online, you would be mistaken. Disinformation may soften an adversary, sowing doubt in the population; cyberattacks may weaken the day-to-day operations of their economy. But if that doesn't achieve your objective, killing their soldiers might. Some of the most rapid exponential changes in conflict have come about on the physical battlefield.

Perhaps the best example is drone technology – pilotless aircraft controlled by soldiers from a remote location. The rise of drones might, at first, seem like a wholly different force to the growth of cyberattacks and disinformation. Viewed more broadly, however, we can see how drone warfare intertwines with and compounds other forms of attack.

War has long been fought simultaneously on multiple fronts: today, disinformation and cyberattacks tenderise an adversary, then a drone attack masticates them. In all cases, the steeply declining prices of weaponry catalyse conflict – drawing ever more actors into a digital or physical battlefield. The result is a more multifaceted form of warfare than ever, fought simultaneously in private forums, in public life and in the streets. It's new wars on steroids.

The story begins in the early 2000s. Until 9/11, drones played a limited role in military exercises. But as the War on Terror got underway, the CIA started to use them to hunt down the leaders of al-Qaeda in Afghanistan, Pakistan and Yemen. And over the next few years, their uses escalated. The US military's Predator drone is now a ubiquitous and controversial force across the Middle East. President Obama used them extensively, launching his first drone strike three days into his presidency in 2009. By the time he left office, the US had launched 542 drone attacks, one for every eight days Obama spent in the Oval Office. Those strikes killed 3,797 people – a little under a tenth of those were civilians.[45]

But this technology is no longer the preserve of the US military. Drones have been getting cheaper, thanks – as ever – to the rapidly changing nature of the underlying technology. The consumer drone industry (the force behind those 'quadcopters' that blight quiet strolls on the beach) has developed rapidly, riding the exponential curve of advances in microchips, batteries, robotics and AI. Starting in the early 2000s, radio-controlled aircraft gradually became easier to use, more reliable and quieter, thanks to the development of lithium-polymer batteries and brushless electric motors. Next came drones loaded with chips, which allowed hobbyists to tool them up with autopilot systems. Then came accelerometers, which test for changes in an object's speed: as these dropped in price, they were integrated into drones. Much of this was down to the development of smartphones. Accelerometers were developing quickly because they were used as smartphone tilt sensors; minuscule Wi-Fi chips and camera sensors were being made more cheaply for smartphone use too.[46]

The result was a precipitous decrease in the price of drone technology. By the beginning of the 2020s, drones were cheap enough to be used in lieu of fireworks for public events – hundreds of little aircraft carrying many-hued lamps, choreographed by programmers. Many rich countries used these advances to build ever more sophisticated, ever more expensive materiel. A 1990s classic of military aerial surveillance, the MQ-1 Predator combat drone, cost $20 million per system unit (about $34 million in 2019 terms);[47] whereas the Global Hawk drone downed by the Iranian military in 2019 cost $130 million.[48] But other countries took advantage of exponential improvements in the core technologies to build drones on the cheap. The Turkish Bayraktar TB2, launched in 2014, is on the market for around $5 million.[49]

Along the way, these cheaper drones brought wholly new actors into military conflicts – battlefield warfare became affordable to any number of terrorist groups and small states. Military drones represent the inverse of traditional flagship military technologies – the ballistic missiles, nuclear weapons, aircraft carriers and stealth fighters that remain the privilege of a handful of rich nations with mature defence industries. Increasingly, anyone can build a drone army. In a 2018 attack, the Kurdistan Workers' Party (PKK) targeted several Turkish government locations with drones that anyone could buy on Amazon for under $300, having packed them with explosives.[50] Then, in 2019, Houthi rebels from Yemen launched drone attacks against the world's largest oil processing facility in Saudi Arabia, stopping about half of the kingdom's output for a day.[51] In the language of military theory, the development of drones has driven an increase in 'asymmetric' conflict – between powerful, well-established actors like nation states, and upstart combatants with fewer resources.

Conflicts between nation states have also been exacerbated by drone warfare. When fighting broke out in the Nagorno-Karabakh region between Armenia and Azerbaijan in late 2020, drones were used on both sides. A regional proxy war involving Russia, Turkey, Israel and some Gulf nations, the conflict became a showcase of the latest drone technology. Thanks to its close links to Turkey and Israel, Azerbaijan was able to unleash a fleet of advanced drones on the battlefield, which ultimately gave it the upper hand. Such drones allowed Azerbaijani forces to penetrate deep into Armenian territory and wreak havoc on their supply lines.[52] The Armenians lost more than 185 tanks and hundreds of other pieces of heavy military equipment – more than half its materiel at the start of the conflict. There was a human cost too. Azerbaijan took to launching kamikaze drones against Armenian soldiers, leading to, in the words of one US-based military analyst, 'very heavy human losses'.[53] This conflict was a turning point, sparking a realisation that the men-and-materiel approach of the twentieth century was vulnerable to low-cost exponential technologies. Within weeks, the UK's Ministry of Defence had turned its attention to buying cheaper armed drones.[54]

But the issue is not just how cheap they are. Drones are more agile than many types of regular weapons. They can be launched from relatively low-key locations and can access hard-to-reach areas. The Chinese military has tested launching swarms from a light tactical vehicle and from a helicopter, for example.[55] As agile reconnaissance tools, meanwhile, drones will improve accuracy, resulting in fewer wasted warheads. Acting as decoys, swarms could even confuse enemy radars and lure their forces away.[56] Such weapons arguably also make illegal forms of warfare more likely, should any nation have the appetite to break international law. Take chemical warfare: instead of carrying large loads of a chemical or biological agent and spraying it en masse, drones can target specific vulnerabilities, such as the lead vehicle of a convoy or the home and garden of a target.[57] In theory, only a rogue state would ever consider such a method; but the increased reliability and precision of chemical warfare may prove a mite too tempting for some nations.

What is most concerning is that the exponential transformation of drones is still in progress – meaning it is hard to guess where this technology will end up. Human-controlled drones are the first step towards 'autonomous systems'. Increasingly, militaries and paramilitaries are looking into weapons that can engage the enemy without human intervention. Autonomy is a continuum, not a binary. The US Tomahawk missile has been in use for over 30 years; it can navigate by itself with no GPS, and maintain accuracies of under a metre.[58] It has some autonomy but is not fully autonomous: the human is still in the loop to 'tell' the computer what to hit and when. However, fully autonomous drones may be possible. And the Israeli-developed drones Harpy and Harop are often cited as autonomous weapons that are already in use. Harpy uses electromagnetic sensors to search for pre-specified targets; its follow-on system, Harop, uses visual and infrared sensors to hunt those targets.

The development of facial recognition and computer vision will further add to the power of such technology on the battlefield. A commercial drone not even intended for military use, the Skydio R1, uses a cutting-edge computer vision system to recognise and track its owner autonomously. Thirteen on-board cameras do real-time mapping, path planning and obstacle avoidance.[59] It sells for less than $2,500. Such drones, with high degrees of autonomy, are more and more commonplace. In 2017, the Stockholm International Peace Research Institute analysed 154 weapons systems with automated-targeting capabilities and concluded that more than one-third could select targets and attack without human intervention.[60]

In the near future, autonomous systems could support warfare by undertaking thousands of complex and coordinated actions in a faster and more agile manner than humans. Communication delays, currently an inevitability as drones are constantly communicating with human-staffed command centres, could be a thing of the past. With synchronised, sensor-equipped drones, each machine carrying a computer on board, this would no longer pose a risk.[61] At its most extreme, this could lead to a fully autonomous weapons system, which could make its own decisions about whether, who, when and how to attack. Humans might roughly outline the battlefield, the objectives, the nature of the adversary and the rules of engagement. But the machine would pull the trigger, deciding whether someone is an enemy or not; whether to fire or not; whether to kill or merely disable them.

One need not have read much sci-fi to know that this feels a little dystopian. All of the current norms and regulations surrounding war are rooted in an assumption that humans are in absolute control. The International Criminal Court is set up to investigate war crimes allegations made against individuals such as politicians or militia leaders.[62] The Nuremberg tribunals affirmed the importance of human responsibility and the culpability of individuals. Legal experts argue there is a lack of clarity surrounding how courts might deal with crimes arising from the use of autonomous or semi-autonomous weapons. There is, in the words of one recent paper on the subject, a 'responsibility gap'.[63] Humans, presumably, are ultimately liable – but which ones? The engineers or the programmers? The generals or the politicians? And what if the machines are autonomous, provided with general directives but not specific missions? In the absence of a Geneva Convention for an age of autonomous warfare, the answers remain unclear.

The exponential world is disordered. More actors have more methods with which to strike larger attack surfaces. In this world, war is cheap – in financial and human terms. Unlike Colonel Raz and his F-16 pilots heading into Iraq, a malicious social media campaign that riles an opponent's citizens puts no soldiers at risk. And the results can be varied, and disorientating: ranging from nuisance-level misinformation to meticulously planned disinformation campaigns; from cyberattacks against power and water infrastructures to drone-borne explosive attacks. Perhaps all foreshadowing a full-blown war.

At the same time, our political order is tending towards instability – it is hard to predict what effect the re-localisation of global society will have on warfare, but it seems unlikely the outcome will be peaceful. Combined with new technology, this trend looks set to turn the Exponential Age into one of apparently never-ending conflict – where any low-level scuffle could escalate into something more disastrous. The mischievous needling of small-scale cyberattacks, the disruptive potential of long-term disinformation campaigns, and savage, co-ordinated cyber warfare are all part of the same ecosystem. And citizens and businesses, as much as politicians and armies, will be on the front line.

What can we do to escape this world of fragmentary, high-tech conflict? Since Thomas Hobbes's *Leviathan*, published in 1651, scholars have often argued that the first duty of the state is to protect its citizens from violence. If states can't prevent their citizens being targeted by Exponential Age weapons, they are failing in their most fundamental obligations. And so it falls to states to resolve the escalating conflicts of the Exponential Age.

There are three ways they can do so. First, states need to strengthen their defences across their entire attack surface – undertrained militaries, insecure office computers, gullible citizens. Second, they must establish new norms surrounding communication and escalation. Third, they must work to reduce the proliferation of attack surfaces and weapons.

Let's take these one by one. Western states' defences against the possibilities of high-tech attacks are often weak. They are like castles without moats – or castles in an age of helicopter gunships that can fly over walls. But with the right investments, they could build moats (or anti-aircraft guns). On one level, that involves a rebalancing of the priorities of the military. It takes the realisation that expensive planes – like America's Lockheed F-35 fighter, which will cost nearly $2 trillion over its life – are no match for cyberattacks: and that cyberattacks are a real threat. And so, in many cases, investment should go elsewhere.

This is something that many nations are gradually realising. In March 2020, the US Department of Defense updated Congress on its aforementioned policy of 'defending forward'.[64] Their strategy involved a more active approach to protecting US national interests in cyberspace, by strengthening the role of Cyber Command within the Department of Defense.[65]

But boosting national defences will not just involve defending state security infrastructure. Much of the digital infrastructure of the twenty-first century is owned and run by private companies. This means that the sites of attack are privately controlled; and it means that identifying, responding to and recovering from an attack will often fall to private corporations. When al-Qaeda operatives attacked New York and

Washington in September 2001, it fell to the US military to respond. When Maersk and Mondelez were hit by cyberattacks almost two decades later, it was corporations that mounted the first defences.

We might boost the security of these nation-sized companies by increasing their public accountability. When it comes to countering cyberattacks, massive companies do not have a seat at any formal decision-making table, nor clear ways to be held responsible for what happens on their platforms. Many of these firms take their security obligations seriously nonetheless. Microsoft, for example, has embraced its newfound role in preventing malicious attacks with an initiative called the 'Digital Geneva Convention', which is designed to protect civilians from cyberattacks. Similar to the Geneva Convention, which offers protections for civilians, prisoners of war and the natural environment during times of conflict, the Digital Geneva Convention aims to engage both governments and private companies as guardians of civilian security. Governments might develop a similar but broader system, laying out the obligations of private companies that operate digital infrastructure.

And then there's the biggest attack surface of all: every single one of us. Any of us could fall for a phishing attack, or accidentally turn our home into part of a botnet, or find ourselves doubting scientific evidence and refusing to take a vaccine. As the retired general Richard Barrons told me, many in the West are 'strategic snowflakes'. 'Our daily life is enormously fragile,' he says.[66] The solution is digital literacy: the set of skills that enables citizens to operate safely in the digital world. An important feature of this is digital hygiene – making sure your passwords aren't 'password', knowing how to clean up old accounts, turning on multi-factor authentication and being able to identify a phishing scam. At the same time, citizens need training on how to function in a world of digital media. That often rests on media literacy, the idea that citizens should think critically about the memes and messages they come across, whether on the internet or in print.

One nation leading the way here is Finland. Just a short ferry ride over the Barents Sea from Estonia, and sharing a land border with

Russia, the Finnish state is perhaps understandably mindful of the perils of misinformation. Here, a partnership across the public and private sectors emphasises a coherent response to disinformation – particularly that of Russian source. The programme, launched in 2015, was developed by the president, Sauli Niinistö.[67] Critical thinking skills form part of high school curricula, as well as later adult education.[68] Children even take classes on how to recognise fake news. A second country leading the way is Taiwan, similarly dwarfed by an often-hostile neighbouring state. Many of Taiwan's policies are geared towards the threat posed by China, and their misinformation strategy is no exception. Taiwan's digital minister, Audrey Tang, told me that they 'teach media competence, meaning that instead of being just readers and viewers of data and journalism, everybody is essentially a producer of data and narratives'.[69] It's a lesson that many in the West would do well to heed.

Stronger defences can make foreign attacks less effective and more expensive. But they aren't sufficient in isolation. Defence is fundamentally reactive – it cannot prevent attacks happening, at least not directly. And so a second area of improvement relates to de-escalation. How can nations make sure they are communicating in a way that reduces the likelihood of attacking each other in the first place? And if attacks do begin, how can leaders stop them spiralling out of control?

In both cases, we can learn from the experience of the Cold War. While there were some fraught moments, the twentieth century did not, fortunately enough, end with nuclear oblivion. And this was in part down to the communication and de-escalation tactics adopted by some of the more sensible officials in Washington and Moscow. For one thing, the two superpowers ultimately learnt to interact effectively. After the Cuban missile crisis in 1962, the United States and Soviet Union established a hotline – the famous 'red telephone' – so the superpowers' leaders could talk to each other directly. It was an important measure in building trust between the two rivals, particularly at times when a small miscommunication could spell disaster. Policymakers might build a similar network of communication fit for the digital age. As NotPetya showed, cyberattacks are messy – they hit their target but also much else besides, travelling rapidly across our global information networks. This ubiquitous digital fallout can sow confusion about what the real target was, and increase the likelihood of escalation. Better lines of communication would mitigate this risk.

There is also a dire need for new rules of engagement and methods of escalation and de-escalation. Cyber conflict remains a young domain, and the rules are evolving at different rates among different adversaries.[70] If nations increasingly 'defend forward' as the US has started to do, the chances of misunderstandings or conflagration increase. Again, the less hawkish policymakers of the Cold War hint at a way forward. By the end of the Cold War there were clear norms and doctrines in place, designed to prevent conflicts escalating accidentally. Clear, unambiguous lines were drawn. For example, NATO's Article 5 was designed to be a hard trigger: any attack on a NATO member was deemed to be an attack on all. Such lines have become blurred during the Exponential Age. It was not until 2019 that NATO started to officially consider whether a serious cyberattack would call for the collective defence enshrined in Article 5.[71]

And then there is non-proliferation. Through formal or informal means, we may be able to make certain forms of attack unacceptable – either by law or by convention. We have pulled this off before. Chemical and biological weapons, banned internationally in 1925, remain rarely used – even by states that are fond of cocking a snook at the rules-based international order. Landmines have become increasingly rare since a treaty of 1997, cluster bombs since a treaty of 2008. In many cases, such treaties work because they compel private companies, such as arms makers, to stop manufacturing certain classes of weapons. This means they cannot fall into the hands of non-state actors, as well as national armies.

What would a set of non-proliferation treaties fit for the digital age look like? We might start by taking on hacking. In 2021, the

CyberPeace Institute called for preventative action against commercially available hacking software – memorably termed 'intrusion as a service'.[72] Here too, a key goal would be to ban private companies from developing malicious software. Take the NSO Group, a private firm which develops software purportedly used as spyware by malicious actors.[73] Their blockbuster product is Pegasus, which allegedly targets individuals' phones and then takes them over. So far, dozens of journalists and activists around the world allege that Pegasus has been used to spy on them.[74]

Looking to the future, we need to develop a legal framework to govern the use of autonomous weapons. Today, only a small minority of countries have advocated a complete ban. But there is a wider consensus that autonomy cannot be limitless – and 'human control' must remain an important aspect of weapons design, operation and governance.[75] Work is needed – and fast – to establish new rules that keep pace with the technological acceleration of our weaponry.

At the time of writing, many of these policies feel out of reach. It is not just rogue states and terrorist groups who are drawn to the methods of exponential warfare. We've seen how the US military came to depend on drone strikes during Barack Obama's presidency. And, as the example of Stuxnet shows, cyberattacks can be a helpful weapon in the arsenal of rich nations. It is hard to call for a moratorium of a particular tactic when you yourself engage in it with aplomb.

But inaction is not an option. Exponential technologies will drive further conflict, both by eroding our international order and making warfare more viable. Today, everything is a potential attack surface, and everyone needs to know how to defend themselves. At the same time, malicious actors have new tools to wage the oldest kind of bloody war in the physical world. Without new systems of defence and de-escalation, the world risks becoming a sprawling, messy battlefield.

The talk of national defence, however, raises further questions about life at home. The primary goal of any sovereign nation is to protect its citizens from violent threats so they can have safe and fulfilling lives. But conflict is only one of the threats to that sense of fulfilment. Other areas of our lives – the relationship between consumer and market, citizen and society – are also being unsettled by exponential technology. And, as we'll discover in the next chapter, we must do more to make sure that the society we inhabit is worth defending.

8

Exponential Citizens

On 8 June 1972, the South Vietnamese air force dropped napalm on the village of Trang Bang, about 35 kilometres to the north-west of Saigon. A nine-year-old girl, Phan Thi Kim Phuc, was hurt in the attack. She suffered third-degree burns on 30 per cent of her body from the unctuous mix of gasoline and aluminium salts that fell on her home town.

A photo of Phuc running away from the attack, naked and crying, would become one of the most famous anti-war photographs ever taken. Nick Ut, the photographer, would win a Pulitzer Prize for the shot in 1973, and it remains a powerful symbol of the brutality and horror of war.

Not according to Facebook, however. Some 44 years later, a journalist at the Norwegian newspaper *Aftenposten* published a story examining the history of war photography. He promoted the story – accompanied by the photo of Phuc – on Facebook. Hours later, Facebook removed the photo for breaching its policies on child pornography. When challenged, Facebook responded by suspending the journalist and issuing a statement that doubled down on their position.[1] 'While we recognise that this photo is iconic, it's difficult to create a distinction between allowing a photograph of a nude child in one instance and not others,' the company said.[2] As the row escalated, Erna Solberg, the prime minister of Norway, posted a copy of the photo herself – only to see her version removed by Facebook too. Only after a public outcry and global media coverage did the firm recant, reversing its earlier decision to censor one of Norway's leading newspapers and the country's premier.[3]

In September 2016, this still counted as a serious controversy for

Facebook. The notion that such a major editorial decision could be made by fiat by some Palo Alto executives was shocking. But within months, the power Facebook wields over our public sphere would become all too apparent – when controversy surrounding Russian interference in the US election via Facebook reached fever pitch. And, over the next four years, Facebook would face ever-greater questions about what – and who – it chose to censor, and what it didn't.

Facebook is the most powerful media platform in the world – the largest such network in most big countries apart from China. Facebook and its subsidiaries are also the very lucrative plaything of Mark Zuckerberg. In the words of one of its co-founders, Zuckerberg's Harvard roommate Chris Hughes, 'Mark's influence is staggering, far beyond that of anyone else in the private sector or in government.'

Not only does Zuckerberg profit massively from the company, he also has near-absolute control over it. 'Facebook's board works more like an advisory committee than an overseer, because Mark controls around 60 percent of voting shares,' says Hughes. 'Mark alone can decide how to configure Facebook's algorithms to determine what people see in their News Feeds, what privacy settings they can use and even which messages get delivered. He sets the rules for how to distinguish violent and incendiary speech from the merely offensive, and he can choose to shut down a competitor by acquiring, blocking or copying it.'[4]

Hughes's analysis of Facebook reveals much about life in the Exponential Age. Throughout this book, we have seen how exponential technologies are disrupting the relationship between old companies and new, employer and worker, global and local, national defences and high-tech attackers. But we haven't yet explored arguably the most important relationship – the one between citizen and society. And, in particular, between citizens and the market.

For centuries, we have accepted that there are some areas where private companies belong – and others where they don't. Buying and selling is generally an efficient way to supply the world with coffee, or books, or technological widgets. But there are some things that cannot be bought and sold. The issue is not whether buying and selling in these arenas would be efficient, but rather about what is ethically justifiable. People's private selves, we often think, shouldn't be on the market. You can't buy a human. You can't buy my liver. You can't buy a friend – or, if you did, people might judge you a little. And then there is the public square, in which we discuss the kind of society we want to inhabit. Democratic institutions like congress or parliament or, in many countries, public service broadcasters – all fall beyond the reach of the market. These binaries have always been blurry – if I buy a newspaper publisher, am I not buying up part of the 'public square'? But we generally accept them nonetheless. There are, in the words of Harvard philosopher Michael Sandel, 'moral limits' to markets.[5]

However, in the Exponential Age, private companies increasingly encroach on these areas – ones that we once thought of as beyond the market's reach. This is the result, once again, of increasing returns to scale. Exponential companies get bigger and bigger, sometimes verging on monopoly. They expand horizontally, at an accelerating rate, into ever-wider sectors of our society. And our democratic norms, embedded in slow-to-adapt institutions, seem unable to keep them in check.

This is a phenomenon with three key features. The first, best exemplified by Facebook's arbitrary controls of conversation as discussed above, is the emergence of new, private rule-makers – whose growing power amounts to the privatisation of the 'public sphere'. This term, first coined by the German philosopher Jürgen Habermas, refers to the arena in which private individuals come together to discuss the needs of society and the laws that govern it. As the name suggests, this is supposed to take place in public. We have long thought that laws should be made by accountable, elected officials; and they should be scrutinised transparently and collectively. Today, however, a handful of private organisations increasingly dominate that public sphere. They, rather than elected parliaments, make the laws that govern our lives. And our public conversation is conducted on a small number of private platforms, rather than in a wide array of newspapers and broadcasters, think tanks and coffee shops.

The second is the encroachment of markets into our private lives. The private sector and our private lives were once very different. We bought and sold our houses, our food, our labour. But there were parts of our lives that we wouldn't sell. The conversations between lovers or within families. Essential, personal information about our health, or our relationships, or our sex lives. Increasingly, private companies monitor – and lay some kind of claim to – all of these. As our conversations have moved online, each of us has come to develop a digital doppelgänger – a more intimate account of who we are than we could ever write ourselves. This takes the form of data. Data about individuals has exploded since the growth of social media and mobile phones. The volume of real-time data generated in the global economy expanded by a factor of 19 between 2010 and 2020 – a doubling just under every two years, according to the analyst firm IDC. In 2020, more than 42 million WhatsApp messages were sent a minute, and nearly 150,000 photos uploaded to Facebook.[6] This data could be used beneficially, to solve our knottiest societal problems. But today it is largely bought and sold for profit – and along the way it is used to profile us, exploit us, get our money.

The third is the privatisation of the way we interact with one another. It is not just our relationships with the public square and our relationships with ourselves that the private sector is coming to dominate. Our personal relationships – the ways we form social bonds, make friends, build communities – are being remoulded by private platforms. Until recently, communities evolved without the pervasive controlling influence of any one private organisation. Corporations might nourish our collegial instincts – as anyone who has ever chatted with a friend in a chain coffee shop can attest – but no one business determined how social ties came about. But today, our communities are often created online, via platforms owned by a handful of oligopolistic companies. Facebook, Twitter, TikTok: each has a massive impact on who we meet and what we think of them. And this may lead to an increasingly divided society, polarised into thousands of groups that barely interact.

In all three realms, there is a shift in where the power lies – out of our hands and into those of a small number of technology executives. The decisions they take will, understandably, be ones that benefit their businesses and themselves. Sometimes they reflect the needs of society as a whole. But this is usually a mere coincidence. And in many cases, they have harmful effects on society.

With time, a gulf emerges between our ideas surrounding the place of markets, and the role they actually play in our lives. And this process will only continue. As companies get bigger, they will encroach ever further into the realms we previously thought of as beyond the market. The result is a fundamental rewiring of the relationship between citizen and society.

In the early days of the internet, a few people figured out that the digital realm might become governed by the technologists who built it – rather than by the consensus and needs of those who used it. The communitarian ideals of the early internet inadvertently handed power to those who knew how to code and could harness the power of the nascent web for their own purposes.

Larry Lessig, a Harvard law professor, was one of the first academics to study the emerging power of the internet's developers in determining what we could and couldn't do. Code – the instructions that make up software in general and the net in particular – could become a constraint on our behaviour, he argued. In much the same way that formal laws, developed constitutionally by legislatures, prescribe our actions, choices in software would govern our behaviour in the digital realm.

His famous essay 'Code is Law', published at the dawn of the new millennium, remains an excellent, prescient text. 'The code regulates,' Lessig writes. 'It implements values, or not. It enables freedoms, or disables them. It protects privacy, or promotes monitoring.' This means that coders have untold power. 'People choose how the code does these things,' Lessig argues. 'People write the code. Thus the choice is not whether people will decide how cyberspace regulates. People – coders – will. The only choice is whether we collectively will have a role in their choice – and thus in determining how these values regulate – or whether collectively we will allow the coders to select our values for us.' In Lessig's view, we did not sufficiently scrutinise the interests of developers – the very people forging the laws of the internet.[7]

This is the first, arguably most corrosive, way the market is stepping into once-uncommodified areas of society. It took a while to become apparent. But Lessig's prediction – that, in the absence of considered government action, tech elites would become the rule-makers of the digital world – has largely come to pass. In fact, these rule-makers are arguably more unaccountable than even Lessig predicted. Because it's not individual coders but rather big and ruthless corporations that determine the laws. In Lessig's day, the best technology companies were largely run by their developers, be they Microsoft or Yahoo!, and this engineering culture called the shots. But since that time, the shape of many of these firms has morphed. They have product managers and policy teams who have a strong hand in shaping the code. And these executives are arguably more focused on commercial maximisation than the motley hoard of 1990s coders ever were. After all, today the prize is much bigger.

The power of coders – and their bosses – has implications for two interrelated parts of our public sphere. Broadly, as Lessig warned, it alters who the lawmakers are. And more specifically, it alters what ideas can get into the public conversation – the tech companies become the censors. Facebook makes the rules; and the primary place you can talk about Facebook's rules is, you guessed it, on Facebook.

It's worth reflecting on these two changes in turn. First up, the growing power of private actors in making the rules about what can be done – and what can't. To see the power of this shift, we must pay a visit to Harvard in the early 2000s. What is now Facebook started as FaceMash, the result of a teenage college student's peccadillo. Mark Zuckerberg and some of his friends built a website to judge the attractiveness of female students. It soon morphed into TheFacebook, an embryonic social network that played up to the dating-and-mating culture of American colleges. Naturally, the early developers decided that it was important to ask for someone's relationship status during registration – and that one of those relationship statuses would be 'It's complicated'. Both of those decisions, small in themselves, made sense in TheFacebook's early context.

But as Facebook became more successful, this dorm-room thinking became the global standard. If you wanted to enjoy what Facebook offered, you had to subscribe to that standard. And Facebook exported these particular sociocultural qualities to societies where such questions were inappropriate, or meaningless. People like my mother, then a septuagenarian of Muslim heritage, were offered the choice of an 'It's complicated' relationship. In her context, and doubtless the contexts of hundreds of millions of other people, it was a nonsensical, perhaps even offensive, question. But the code had become law.

If this example sounds trivial, many others are anything but. Consider the Covid-19 pandemic. Tackling a highly contagious, airborne virus with a long incubation period was always going to be a tall order. Given the speed with which it was spreading, tracking the virus would be hard to do manually. Fortunately enough, many of us now carry smartphones that know where we are. They could be used to 'contact trace' – that is, work out who potential virus carriers had interacted with and when. The trouble was, virtually every smartphone in the world runs software made by Apple or Google. And these firms decided, without government mandate, to step into the fray. They updated their operating systems to enable some underlying components, which made building contact-tracing apps easier. According to Apple, both firms made design decisions that 'privacy, transparency and consent [would be] of utmost importance'.[8] The tech giants' software updates became a contributory factor in allowing governments to build more effective contact-tracing apps. In this case, few could impeach the decision made by Apple and Google – it helped slow the spread of the virus. But it also underlined that it was Apple and Google who determined what governments and scientists could – and could not – do. The code had become law.

To say this was an important decision is an understatement. Apple and Google helped enable tools to support contact tracing, but did so on their terms – prioritising user privacy over government surveillance. This may have been a good call, but should decisions like this be taking place in the private realm? Usually, laws are made by elected officials. Now, we have a system of arbitrary power without any clear process of accountability. In the words of the internet activist Rebecca MacKinnon, the 'consent of the networked' – that is, every one of us – is being weakened.

And then there's the question of censorship. This is, arguably, the key consequence of the emergence of the new rule-makers. Today, the question of what makes it into the public sphere – and what doesn't – is dominated by an ever-diminishing number of technology executives.

The net is, on one level, a democratising force. An internet connection gives you the ability to talk back. Before the internet was widespread, media was controlled by a small number of TV, radio and newspaper outlets. They talked, you listened. Participating in the public conversation meant chatting with friends at drinks after work, or writing a letter to the editor and hoping it was published. Whereas today – thanks to email, blogs and above all instant messaging and social networks – members of the public can express themselves publicly, and in larger numbers than ever before. The internet has set our public sphere free.

And yet, paradoxically, that same public sphere is increasingly concentrated in a few people's hands. Facebook and Twitter are the key spaces for political conversation, community gossip, blathering about the news. The internet in general, and social networks in particular, are the only places where many conversations can take place – because it is the place we all are. Once again, the network effect is all-important – the more people use these spaces, the more these spaces become the only ones in which discussion can take place. Their scale means that these platforms can censor (or promote) what they choose. And their motives are often what is good for business, not what is good for society.

In some moments, the problem is censorship – think of the photo of Phan Thi Kim Phuc, expunged from Facebook. In others, the problem is a lack of censorship. In 2018, several Indians were murdered by violent mobs who had been agitated by viral messages spread on WhatsApp.[9] Facebook – WhatsApp's owners – stood by as the situation escalated into violence. But in all cases, the root problem is that an unaccountable company decides what ideas get circulated, and how. In the wake of the killings in India, Facebook limited the number of times messages could be forwarded. Initially, the firm took moderate steps to do so in India alone.[10] Eventually, more extensive measures were introduced worldwide.[11] Over the course of two years, what you or I could do on WhatsApp changed. And so too did what we could say to each other. The basic mechanism through which ideas enter the public sphere was transformed.

In this case, Facebook arguably made the right call in the end. In others, their response has been more problematic. For several months in 2018, army generals in Myanmar were fomenting violence against the Rohingya Muslim minority using Facebook as their primary communication tool. Facebook acknowledged it was 'a means for those seeking to spread hate and cause harm' in what has since been described as the ethnic cleansing of the Rohingya minority.[12] (Facebook would later reject a request to share data that could have helped hold the Myanmar military accountable at the International Court of Justice.[13]) Facebook also allowed imminent threats to American citizens to stay live on its site in the face of dozens of reports from users. A day before Kyle Rittenhouse shot and killed two unarmed protestors in Kenosha, Wisconsin, a Facebook page he had visited was urging people to take up arms. Facebook users reported it more than 455 times, but the company left it visible.[14]

Many platforms make an apparently arbitrary distinction between what is acceptable to say and what isn't. The banning of Donald Trump's social media accounts in January 2021, not just by Facebook but also by Twitter, YouTube, Twitch and others, demonstrated the randomness of this line. Trump, it seemed, had gone beyond the pale when he allegedly encouraged the storming of the US Capitol building. But was this necessarily more unforgivable than his earlier behaviour? Over the previous decade, Trump had used social media to promote a racist conspiracy theory about Barack Obama's birth certificate, amplify tweets from far-right and white supremacist accounts, and falsely allege widespread voter fraud in a series of fair elections – to name only a handful of many, many examples. The new technological platforms had bent themselves into contortions over how to deal with Trump and with abusive content in general. When the social media bans came in a 36-hour period during the transition between Trump's election defeat and Joe Biden's inauguration, it raised some crucial questions: Why now? Why this?

The apparent inexplicability of Facebook's decisions – whereby Trump's Capitol tweets sparked a ban but a decade of misinformation didn't; and where the photo of Phan Thi Kim Phuc is censored but anti-Rohingya hate speech isn't – reveal the power exponential platforms hold. Our public conversation now occurs within systems they control – systems governed by their terms of service and their commercial incentives.

The issue here is less whether the digital platforms are making the right calls, and more whether we want these questions to be handled by private companies in the first place. Some corporations are now experimenting with new approaches – in May 2020, Facebook created an 'Oversight Board' of experts on journalism, human rights, law and politics, with the power to scrutinise and overturn the company's decisions on censorship. It is not a bad start, but the board still seems a mere shim, introduced to curb Facebook's worst excesses. With no clear guidance from legitimate policymakers, Exponential Age companies have made up the rules themselves.

As Facebook's top public relations executive – and former deputy prime minister of the UK – Nick Clegg wrote, 'It would clearly be better if these decisions were made according to frameworks agreed by democratically accountable lawmakers. But in the absence of such laws, there are decisions that need to be made in real time.'[15]

In the Exponential Age, each of us exists in more than just flesh. We are all reflected in a set of numbers, constructed from our digital trails around the internet – our data selves. This data provides a more detailed account of our likes and dislikes, hopes and fears, than many of us can offer about ourselves. And this data, too, has been privatised.

This is the second way in which markets have come to encroach on our lives. Businesses, most notably those that use data to sell advertising, have become addicted to our digital trails. And with that, they have taken our private lives into the private sector.

Carissa Véliz, a philosopher at Oxford University, argues that this is fundamentally an issue of power. Privacy, she says, is power. Knowing personal details about someone is a very particular kind of power, because it can be transformed into economic, political and other kinds of power. 'Power over others' privacy is the quintessential kind of power in the digital age,' she writes.[16]

Our private data can be used for any number of purposes – to manipulate our behaviour, to profile us, to target us with particular messages. In this sense, the European Union's law on data, the General Data Protection Regulation, is telling. The GDPR considers each of us to be 'data subjects' – that is, someone about whom someone else holds data. But this term might also have another meaning – that we have all become subjects of data. Data is certainly used to profit from us; it is often used to track us; and it might be used to control us.

The ways companies track our behaviour and generate data from it are becoming ever more complicated. When, in the mid-1990s, the internet first became a destination for consumers rather than just academics, user tracking was very limited in scope and application. When I launched *The Guardian*'s newspaper's first websites in 1996, we used primordial tracking to understand how our pages were being accessed. It was so naive: we would count which pages were viewed and the IP address the visitor was coming from. These were the basic 'log files' any web server would ordinarily keep – the information an administrator would need to debug or improve a system. It wasn't used to predict user behaviour or put visitors into cohorts that would be passed on to advertisers. Nor could it have been.

But even this simplistic tracking imposed too much of a strain on our then state-of-the-art servers. They would slow to a crawl. Until we got more capacity, I used to turn the logging off at peak hours of the day. The machines couldn't keep up. And turning off this logging had no commercial impact – it was just a minor technical nuisance for me in my role as server administrator.

That all began to change with Google. When the company introduced its AdWords advertising system in 2000, it triggered a gold rush to gather more and more data. AdWords was a simple tool that allowed businesses to advertise against specific search terms. But it transformed Google's fortunes. The company's revenues leapt from $220,000 in the year before AdWords to $86 million the year after – allowing Google's executives to successfully take the company public just three years after launch.[17] But for AdWords to work, it needed data about what people were searching for, where they were, and more. As Google discovered the profits to be found in the advertising market, it also found the incredible value of gathering data about individual behaviour. Every piece of information, at least in theory, let them make more money.

Where Google led, thousands of other companies followed. Data, they realised, was valuable. And that led to innovative and often rather unscrupulous ways of sucking it out of us. It was miles from my experience at *The Guardian*. Within a couple of decades, a user visiting a few websites might be tracked by hundreds of organisations.[18] Each tracker gathers as much information as it can: your location, IP address, mouse movements, what you searched for, what browser you used, the screen resolution of your computer, and dozens of other data points.

Facebook is, as so often, an illustrative example. When Brian Chen, a reporter at the *New York Times,* downloaded all of the data that Facebook kept on him, he found at least 500 advertisers had some of his personal info, including his contact information and address, even if he didn't know what they were selling or had never heard of them before.[19] Some of that data might not have even been useful for an advertiser. Then again, it might have been. The declining price of storage means it is often easier to just store information than figure out whether or not you need it. Either way, in practice, almost all of the data created on Facebook could be used for the company's purposes. Facebook has struck agreements with at least 60 device makers – including Samsung, Apple and Microsoft – and some of these deals have allowed access to data about people's friends, even without their consent or knowledge.[20]

As a result, we live in a world that is awash with data. The data sphere grows by about 18 million gigabytes per minute. It is a bafflingly huge number – so baffling that no amount of fiddling with it makes it much more comprehensible. Is it easier to say that digital technologies create about 2.4 megabytes of data per human every minute of the day? Perhaps not.

Some of this data is created by machines and sensors, but much is created by us. And, as new exponential technologies emerge, they will generate more data about us than ever before. Health data – from our genes, illnesses and lifestyle – will soon be converted into numbers. Novel technologies such as brain-computing interfaces could pluck neural activity from our heads even before it forms into thought. Wider use of sensors in our homes and environments will gather more clues about how we live. And the relentless march of processing power will further extend the ability of companies to build our data doppelgängers.

But we shouldn't assume that the collection of data is inherently bad. It has many beneficial effects. The more data we have, the more accurate the models we use to make sense of the world become. It can be used to predict whether a population is at risk of disease, or whether an economy is healthy, or whether a business is on track for success. To pick just one among innumerable examples, consider consumer credit bureaux. These are an essential way of determining a consumer's financial health. They harvest data from banks, credit card companies and utilities to build up a financial profile of each of us: our credit score. This data is key in helping us get a mortgage or car loan – without it, lenders are chary to lend. Countries without credit bureaux have much less well-developed consumer financial services than those that do. Data, in this case, should help to broaden access to financial services.

But there's a catch. Is this data gathered for our benefit, or the benefit of the corporations doing the collecting? In the case of credit data, American lawmakers from both parties are not so sure. Credit bureaux often hold inaccurate data, resulting in unfair credit scores. The businesses have also suffered significant data breaches, exposing personal information to the black market.[21] These flaws are the other edge of data's sword: when the information gathered about us works against us.

Such risks extend well beyond web surfing and credit data. Take 23andMe. I developed a fascination with the company soon after it was founded in 2006. Named after the 23 pairs of chromosomes in a human cell, it offers low-cost gene sequencing – giving you the chance to view the source code that makes you into you. It tells you things about yourself that you know, and things you don't. I learnt that my genome suggests that I have wet earwax (I do) and that I'm likely to wake at 7.51 a.m. (actually, I wake up much earlier). It turns out it's possible one of my ancestors was a Viking. I also learnt about my risk of developing diseases like Alzheimer's and macular degeneration.

Fifteen years on, and 23andMe has a huge set of genomic data. Such a data set is incredibly valuable, and may prove helpful in identifying the genetic drivers of disease. Millions of people have used 23andMe's services and have shared their genetic information, and more than 80 per cent of them have allowed their genomic data to be used for research. But 23andMe don't always keep this data to themselves. They can sell it – and they have. In 2018, 23andMe shared the data of consenting customers with pharma giant GlaxoSmithKline for $300 million. Once that data was shared, it was no longer under the purview of 23andMe. It could be shared again and again. And the firm might change its policies; people's genomes might be used in ways they are not comfortable with. In the US, much healthcare data is protected by a 1996 law called HIPAA, which places strenuous requirements on firms handling healthcare data – but genetic testing of the type that 23andMe and others undertake is not covered by HIPAA or similar laws. Rather, it is covered by their own set of guidelines: a nice gesture, but not worth much. Voluntary guidelines have no enforcement mechanism.[22]

At the most extreme end of the spectrum, our personal data can be used to stigmatise and demographically profile us. One example is China's so-called social credit system. In some respects, these systems are relatively benign. Economists often emphasise that the Chinese economy suffers from a 'crisis of trust'.[23] It lacks long-held and consistent business practices. Businesses' affairs were, to put it mildly, disrupted by the 1949 revolution, the Cultural Revolution of the 1960s and 1970s, and the economic liberalisation of the 1980s and beyond. That disruption led to a reluctance on the part of businesses to lend and take risks – and, in turn, a lack of independent credit bureaux. In this context, social credit scores almost make sense. They give everyone a score based on whether they are law-abiding, credit-worthy and a good citizen. It is all made possible by masses of data; and it will, according to researchers, help overcome the crisis of trust.[24] Yet the inherently dystopian implications of a digitally generated taxonomy of citizens is hard to ignore. Recent iterations of the social credit system reportedly put cheating in exams on the list of proscribed activities that could harm your score.

At the time of writing, the democratic world remains blissfully social-credit-system free. However, companies do still use data to profile us – often in ethically dubious ways. Every time you take out a loan, or buy an insurance plan, the terms will be determined by data about similar consumers – and the more sophisticated our data becomes, the more dependent on these methods insurers and banks will be. But this creates a problem: we are all being profiled on the basis not just of our behaviour, but also of our demography. And demographic profiling leads, inexorably, to prejudice. Racial bias in particular tends to seep into data-informed decisions. To take just one example, a study in the US found that an algorithm used by hospitals to allocate care to patients was less likely to refer black patients who were just as sick as white patients.[25] The data the algorithm was trained on reflected racial biases; and so the algorithm behaved like a racist.

As the market inveigles its way ever further into the private sphere, with companies taking personal data about our innermost selves and selling it for profit, these problems will only increase. Our private information is sold back to us, then used to profile and manipulate us. In the Exponential Age, what gets measured gets managed. And in a data economy, we are the ones being measured.

We're often told that opposites attract, but in sociological terms that is verifiably false. One of the most common patterns in life is that we hang out with people like us. Whether it is because of socioeconomic status, ethnicity, a sports team or opinions, we tend to group with people who are more similar than dissimilar.

There are good reasons for this trait, which is called 'homophily'. Try sitting in the home stands of a football derby and supporting the visitors. Or working with people who have completely incompatible views with yours. Or making friends with people with whom you have nothing in common. Homophily makes life easier.

But we have also long known that homophily shouldn't be allowed to get out of control. An atomised society can lead to democratic sclerosis: the more divided we are, the harder it becomes to reach consensus decisions and govern effectively, and the less trust citizens have in political institutions. Eventually, homophily can lead to a kind of social breakdown – studies have uncovered that in some countries, including Turkey and the

United States, people have a growing reluctance to live near other people who support different political parties. One recent analysis of 11 highly divided polities uses the phrase 'pernicious polarization' to refer to the phenomenon of polarisation eroding the ability of a democracy to function.[26]

This poses a problem for Exponential Age citizens. We've already seen how the public conversation and our private lives are being subordinated to the logic of the market. But there's a third aspect that is also being marketised – the way we meet and form bonds with one another. Increasingly, communities are formed online, under the aegis of a handful of digital giants. And when our social bonds are formed solely through Exponential Age platforms, homophily is at risk of spiralling out of control.

Homophily itself is nothing new. It has long been present in our lives – the first study on the process, from 1954, found that racial and religious groups tended to cluster together within housing developments.[27] However, the technologies of the Exponential Age have made a common but formerly low-level feature of our social lives ubiquitous. It is built into the very logic of online networks.

Consider the way people make friends online. Social networks endeavour to link us with people – the 'people you may know' who pop up on the side of your Facebook feed. These people are those with common traits: they went to the same university, or love the same soft rock band, or share your friends. As a consequence, social networks are more homophilous and clustered than the rest of society. And they are more cliquey, ultimately, for business reasons. People who are connected with like-minded people online seem to use social networks more. And groups with similar interests are more useful to advertisers: similar types of people tend to buy similar products, and that means they are easier to sell to. Thus there is a commercial incentive to group similar people into progressively more precise, and discrete, segments.

In practice, however, these methods can end up pushing people into ever more isolated and extreme groups. On services like YouTube and

Facebook, recommendation algorithms are designed to get increasingly bored users to stay on a website.[28] The best way to do that, the algorithms have discovered, is to promote progressively more borderline and extreme material. As the sociologist Zeynep Tufekci puts it: 'Videos about vegetarianism led to videos about veganism. Videos about jogging led to videos about running ultramarathons. It seems as if you are never "hard core" enough for YouTube's recommendation algorithm.' For this reason, Tufekci argues, 'YouTube may be one of the most powerful radicalizing instruments of the 21st century.'[29] And she could well be right. A large-scale study of YouTube analysed more than 330,000 videos and 72 million comments across 349 channels, and found that users consistently migrate from milder to more extreme content.[30] Users who consume extreme content used to consume milder videos – but eventually found themselves following 'radicalisation pathways' that took them to ever more hardcore sections of the internet.

At their most extreme, these pathways can lead people to violence. A report from 2018 set out to investigate whether Facebook was facilitating ISIS propaganda, and showed how important social networks have been in recruiting new members to the terrorist group. It is a study in homophily: people are gradually pulled from the mainstream of political life into more and more extreme communities. Befriending one ISIS supporter leads Facebook to recommend a dozen other pro-ISIS accounts. 'Facebook, in their desire to connect as many people as possible, have inadvertently created a system which helps connect extremists and terrorists,' concluded the researchers, Gregory Waters and Robert Postings.[31]

During their study, Waters and Postings witnessed a Facebook-enabled recruitment – watching a university student in New York go from claiming he had no clear religion to becoming an ISIS supporter, all in under six months. Facebook's own research corroborated these findings. In the words of a leaked memo from Facebook vice president Andrew Bosworth, written a year before Waters and Postings's paper, 'Maybe someone dies in a terrorist attack coordinated on our tools. And still we connect people.'[32]

In recent years, Facebook has come to recognise the scale of the issue. Starting in 2017, the company launched an initiative called 'integrity teams' that endeavours to tackle the problem. What Facebook found as it dug into its own service was dire. Right-wing and racist extremism was rife in many of Facebook's major markets. And the platform's design was making it worse: 64 per cent of people who joined extremist groups joined 'due to [Facebook's] recommendation tools'. In 2018, an internal presentation concluded that the algorithms exploit 'the human brain's attraction to divisiveness', noting that Facebook would present 'more and more divisive content in an effort to gain user attention & increase time on the platform'.[33] It was what many outsiders had seen first-hand, and was now validated by the firm itself.

As new exponential technologies emerge, they look set to make the problem of homophily worse. Genomic Prediction is a company that can help parents going through IVF select healthier embryos. As gene technologies become more widespread, the taboos around their use may reduce. Cliquishness might enter our germ line: we may end up picking children who look like us, or think like us, before they are even born.

Such shifts would lead to a fundamental erosion of the ideals that undergird a functioning society – the shared experiences, values and objectives that bind us together. Homophily stratifies our community. It atomises us into smaller, more disparate, more extreme groups. It builds walls, not bridges.

According to James Boyle, a law professor at Duke University, ours is the era of the 'second enclosure'. The first enclosure largely took place between the fifteenth and nineteenth centuries in Europe, as lands once held in common – fields used for grazing cattle, forests used to forage for wood – were taken into the hands of private owners through a series of enclosure acts. Many economists argue the enclosure movement laid the ground for the industrial revolution, allowing farmland to be used more productively. But it was a drawn-out and often violent process – one that destroyed the social fabric that had held together village life. In the words of one leading economic historian, enclosure meant the rich 'literally robbing the poor of their share of the common'.[34]

In the age of the second enclosure, we once again witness the commodification of new spheres of our lives. 'Things that were formerly thought of as either common property or uncommodifiable are being covered with new, or newly extended, property rights,' writes Boyle.[35] The laws that define our lives are made by private companies; our public conversation takes place on private platforms. The most intimate details of our private lives are bought and sold for profit. And the ways we meet people and forge social bonds are dictated by the proclivities of giant corporations.

But what can we do about it? Within this newly enclosed space, we are at the mercy of the decisions of corporations – it can feel impossible to hold them to account. At the same time, the data economies of the Exponential Age are radicalising us and dividing us – limiting our ability to be good citizens. The two form a pincer movement, lessening our chance to uphold the democratic institutions that allow society to function.

We can all – big companies included – agree that we need new rules. Even Nick Clegg, Facebook's head of PR, argues that the boundaries for what major technology platforms can do must be decided through democratic politics. This will increasingly be the case as corporations come to monopolise discussions on the internet. If we think these companies are harming us, we need to tell them how – and what we want them to do differently. If democratic processes do not determine the rules, then corporations building technologies will. The code will become law.

Today, however, we can begin to sketch out what form these

alternative, more democratic, norms would look like. Our goal here shouldn't be to work out the specific policies that Facebook, or Twitch, or 23andMe should adopt – the individual decisions will vary from case to case. But we can identify broad principles to underpin the relationship between citizens and the market.

First, transparency. The systems that determine how content flows across digital networks – what gets censored, what gets boosted by the algorithm – must be easier to scrutinise. Today's complex exponential platforms are opaque: their inner workings are hidden and their decisions are largely cloaked in corporate pabulum. Greater transparency would allow us to observe how decisions by tech elites are made, and identify what effect they have on us.

Consider what this might mean for content moderation. Silencing the prime minister of Norway over war reportage but allowing the American president to rant on for four years (before abruptly silencing him too) reveals just how opaque these rules have become. Imagine if, instead, we had clear and explicit rules on what forms of speech are deemed acceptable by social platforms, and what forms aren't. This is not an easy line to draw. The more horrific behaviours – such as the live-streaming of the massacre at the Al Noor Mosque in Christchurch, New Zealand, in 2019 – are, in many ways, easier to deal with. Thornier issues arise from online comments that verge on hate speech but might just fall short, or articles that almost seem to incite violence but not quite. At the moment, however, we can't even have a public conversation about where these red lines are – the rules are made in private.

We also need external parties to be given oversight of the outcomes of that process, such as the operation of the algorithms that shape what we do and don't see on digital platforms. When two Boeing 737 MAX aeroplanes crashed in quick succession in 2018–19, Boeing was forced to redesign key aspects of the aircraft. The 737 MAX was not allowed to fly again until it had passed a number of safety inspections. Boeing did not decide when the revamped 737 MAX was safe to fly; the Federal Aviation Administration did. We might learn from this example when thinking about our digital infrastructure. The potential harms brought about by digital networks' algorithms, and their wider processes of deliberation and judgement, need regular inspection – and not behind the closed doors of the companies themselves. This approach might sound radical, but it is no more than already exists in many other sectors. Until 2019, when Donald Trump withdrew America from the INF treaty, Russia and the United States periodically inspected one another's nuclear weapons systems.[36] In the finance industry, bankers regularly run 'stress tests' to identify potential problems with their balance sheets, at the behest of central banks.

What if the decisions that the digital superstars make are legal, but as citizens or users we find them uncomfortable? After all, we may all have different boundaries around what is and isn't an appropriate use of our data – and so the law is of limited use. This is where our second principle comes in: interoperability. We met this concept earlier – the notion that different computer systems should be compatible, and that we should be able to carry data from one to the other. It would mean that if you were using LinkedIn, you would be able to access and read your friends' posts on Instagram or Twitter. Earlier, we discussed how this could help limit the growth of superstar companies. But it can help us as citizens too. It would mean that we would each have more choice. If we were uncomfortable with the way one company was using our data, we could switch to another – while still being able to act in the digital social space. It would force competing platforms to behave better.

The dominant social networks are no strangers to interoperability. They are all built on the most impressive interoperable system we know of: the internet itself. These large firms are increasingly closed and walled off, but it wasn't always this way. Until about 2016, digital platforms were rather more open than they are today. If you wanted to access your Facebook posts or your Twitter timeline, you could use those firm's own products – or you could use them via other services. My own start-up, PeerIndex, could read a user's data from Twitter,

Facebook, LinkedIn and other social networks. While we needed the permission of the user to do this, we didn't need the permission of the social networks themselves.

This approach was facilitated by social media platforms: they largely ran open interfaces (called APIs), which allowed other companies to access their systems. FriendFeed was one such product: it let you look at and post to dozens of different social networks, including Facebook, from a single application. Mark Zuckerberg liked FriendFeed so much that in 2010 he bought the company and subsequently promoted its founder to be the most senior technologist at Facebook.[37] After that acquisition, things started to change. As network effects took hold, clear winners emerged. LinkedIn for business users, Facebook for friends, Twitter for news addicts, Instagram for pictures. Openness played second fiddle to competitiveness. And so the dominant players suffocated interoperability, making their systems deliberately incompatible.

Zuckerberg and his rivals, perhaps, had realised that if you allowed users on one network to talk to those on another network, you handed power to them, both as consumers and citizens, at the expense of the social networks. Interoperability redresses the power imbalance between large platform companies and individuals – if I can leave, say, LinkedIn without losing the ability to post a job ad, then the site's power over me is much more limited. Interoperability is kryptonite to the winner-take-all power of network effects. In an interoperable world, I can stop using a particular provider but continue to reach all of my friends still using the original service.

This type of open interchange is certainly technologically possible, if not completely straightforward. As Sinan Aral, an MIT professor who has studied social networks since the early 2000s, told me, networks 'all have the same messaging formats now anyway: textual messages, video messages and stories'.[38] To make this shift a reality, though, governments would need to act decisively. Nations could mandate interoperability, especially for services that reach a certain size – perhaps 10–15 per cent market share. By waiting until a company had grown to this scale, it would allow founders to innovate and explore markets – while signalling that future success will bring additional civic responsibilities. And the principle need not be limited to social networks. Imagine if a gig worker could bid for jobs across dozens of relevant platforms at once and carry the track-record they have earned from one service over to another. Or in the case of healthcare, where a patient could access their medical records independent of the provider behind them. In each case, power would slowly shift to the person and away from the platform.

Interoperability unshackles our data from any one service, and helps diminish the control the current and future tech giants have over us all. But it does little to address how we should use the data itself. This is an area of vibrant debate. I am not the first person to express concern about the privatisation of our digital doppelgängers and the implications it has for our privacy. It is a topic being discussed by everyone from Ivy League academics to A-list celebrities.

One approach is to embrace the notion of data as property. As the pop star will.i.am wrote in *The Economist* in 2019, 'The ability for people to own and control their data should be considered a central human value. The data itself should be treated like property and people should be fairly compensated for it.'[39] The argument is appealing. By putting a price on our data, we could create a more efficient market for the digital labour we all undertake. Well-known figures ranging from Tim Berners-Lee, who invented the World Wide Web, to Alexandria Ocasio-Cortez, the Democratic congresswoman, have pushed the idea that pricing and compensating us for our personal data would help reform the data economy.

But this would be, in my view, a mistake. As we've seen in this chapter, the creeping monetisation of private life is the problem. Do we want to further marketise our experiences on exponential platforms – to turn our messages to friends and family into a commodity, and allow Facebook to give it a price tag? You might also be disappointed to learn how much you would actually earn from this approach. Facebook made about $7 in profit for every user in 2019. If each user was to take half of the firm's profit, that would net them the price of a coffee. A year. The truth is, an individual's data is worth little – sorry, Mr i.am, even yours.

We don't need to be able to sell our data. We need new rights for our data selves. This is the third principle that would rebalance the relationship between market and citizen: control. We each deserve authority and oversight over our data.

Martin Tisné, of the charity and think tank Luminate, calls for three inviolable data rights: that people be secure against unreasonable surveillance; that no one should have their behaviour surreptiously manipulated by using data; and that no one should be unfairly discriminated against on the basis of data.[40] How, precisely, to enshrine these rights is up for debate.

We could try a digital bill of rights. In the vein of America's Bill of Rights, or France's Declaration of Rights of Man and of the Citizen, a bill of digital rights would legally protect data subjects from the arbitrary will of corporations and governments. But a declaration of rights is a fundamentally reactive mechanism. While it would protect us from overbearing businesses and states, it would do little to entitle us to the true benefits of data. If we're serious about rewiring the relationship between citizens and the market, we need to share the benefits data can bring more widely.

Individually, data – however well protected – isn't that interesting. It is in aggregate that it becomes useful. Data is the thousands of commuters creating information about congestion, which powers apps that help us navigate cities. Data is the massive surveys of patients' reactions to drugs, which let us work out whether new treatments are effective. And data is the shared information from mobile phones that allows us to create accurate weather forecasts. Such pools of data contribute to the common good.

There is one way of organising this data that enhances its benefits without exacerbating the problems of data misuse that plague

Exponential Age firms. But it requires a fundamental reappraisal of the relationship between citizen and society, public and private. One effect of the emergence of free-market economic doctrine has been the emergence of a crude binary. Today, we tend to divide the economy into two camps: the market and the state. In the Friedmanite world view, the state is usually an incompetent, bloated actor. Even on the left, though, the basic logic of the market/state binary is accepted – progressives just have a more optimistic view of government. The trouble is, this dichotomy is flawed. It conceals other, often more productive, ways of organising our lives.

One contemporary of Friedman's, Garrett Hardin, became famous for a 1968 essay called 'The Tragedy of the Commons', which mournfully argues that any resource that is freely available will be overused.[41] The commons are the things that are shared communally – the fish in international waters, or a forest that is shared between a community. And, according to Hardin, they are ripe for over-exploitation. People will act in their own private interest, at the expense of the needs of society – and so the fish will go extinct, the forest will be chopped down. Hardin's argument was catnip to free marketers, implying as it does that the only viable ways to allocate resources are under the auspices of an overbearing state (bad) or through the allocative efficiency of the market (good). Thus the commons were out of vogue.

But Hardin was wrong. The political economist Elinor Ostrom won a Nobel Prize in 2009 for her work demonstrating that the commons – our shared resources – can be well-stewarded without the need to privatise them, or, indeed, nationalise them. In Ostrom's view, the commons are vibrant and self-governing, a complex system that can thrive because it is governed by a patchwork of formal and informal norms. In real commons, such as grazing lands, shepherds would talk to each other to ensure fields were not overgrazed. The commons, it turns out, represent a way of organising and managing resources for the benefit of the people who use them.[42]

And so to the principles of transparency, interoperability and rights, we must add a fourth: commonality. In the Exponential Age we need to create a raft of new commons – which will prevent the creeping power of the private sector in every sphere of our lives. And thankfully, commons are made ever more possible by exponential technology. Digital, intangible assets are unusually prone to management by the commons – they are accessible from around the world, and so can be drawn upon and managed by well-meaning people from anywhere on earth. Unlike fish in an ocean or trees in a forest, digital resources do not get used up. In fact, because of the data network effect, and increasing returns to scale, the more that certain types of information resources get used, the more of those resources there are.

Imagine a world where data resided in data commons. Certain types of information about our lives would be collectively held, in a trusted environment, and used for our benefit. Some data commons are already in existence. The UK Biobank makes excellent use of the millions of data points about our genes that are created each year. It collates the medical and genomic data of 500,000 people. Each has consented to researchers using their data to better understand and prevent serious disease. And any legitimate researcher can access this resource. In fact by 2020, more than 20,000 researchers were doing so. The goal of the UK Biobank is to enable novel scientific discovery with the consent of those who have contributed their data, rather than merely to maximise a commercial return. It is a sharp contrast to how data is often currently managed: hoarded by one party and auctioned off to a small number of well-heeled bidders.

Commons approaches can also be used not merely to share assets but to produce them. If you've used the internet, you have taken advantage of a system built using a commons. The open-source software that powers the engines of web servers and other key internet services is one. Loosely connected volunteers have put down millions upon millions of lines of code, building projects such as Linux OS, the Apache HTTP Server, Mozilla Firefox and the Brave browser. No one owns open-source projects; there are no managers in charge; and no one can prevent anyone from contributing or spinning up an alternative project.

In these digital commons, power doesn't lie in a single place – rather, projects are managed collectively by the volunteers working on them. The best example of such a peer-based commons is Wikipedia: it is visited by more than a billion of us every month, and edited through the collective goodwill of more than 120,000 volunteers. Wikipedia thrives on each participant's commitment to a particular approach to information, a shared sense of mission, and trust that conflicts will be resolved through community-led arbitration. These examples, far from being a tragedy of the commons, bring enormous social benefits. We can think of them, in the words of legal scholar Carol Rose, as a 'comedy of the commons'.[43]

In many cases, commons-based projects have proven to be more successful than commercial ones. And as more domains of the world fall to exponential technologies, the potential of moving industrial activities to the commons increases.

Between them, these four prescriptions – transparency, interoperability, rights and commons – amount to a wholly new way of thinking about the relationship between citizen and market. They respond to the basic problem of modern society: that digital infrastructure undergirds much of our personal, political and public lives. If this infrastructure is exclusively owned and operated by private firms, we will increasingly cease to be fully fledged citizens. We will be mere consumers.

At heart, all four approaches are about limiting unchecked power. Arbitrary power in the hands of anyone has always been a problem. It is a fundamental tenet of liberal democracy – and of any system of good governance – that power should come with accountability, and that those in power should be responsive to the needs of their community. Companies can be as well-meaning as you like and we would still run into the problems that arise from unaccountable power.

This is just as true in the Exponential Age as ever before. Today, new technologies hand untold influence to a handful of companies.

And so these questions of governance and representation become even more urgent. We're in a race not just to limit the new rule-makers' influence, but also to return power to somewhere it will be exercised openly – to put exponential citizens, acting in their collective interest, in control.

Conclusion: Abundance and Equity

On 6 January 2020, a University of Sydney virologist called Edward Holmes posted a short statement on a website called Virological, on behalf of researchers at Fudan University. The site bills itself as a discussion forum for the analysis of viral molecular evolution and epidemiology. It's a niche area, and the site attracts perhaps a few dozen users a month. Holmes's message was functional and unemotive: 'The Shanghai Public Health Clinical Center [and collaborators] is releasing a coronavirus genome from a case of respiratory disease from the Wuhan outbreak. The sequence has been deposited on GenBank and will be released as soon as possible.'[1]

Virologists read the message and hopped over to GenBank, an open-access repository of genome sequences run by America's National Institutes of Health. This strange pneumonia had only been identified a few weeks earlier, originally in a crowded seafood market in Wuhan. Not even a dozen people had died from it. But soon, scientists around the world were on the case. Many started to download the sequence to explore it: as if by teleportation, the code of the virus was transported to computers across the globe.

As the virus spread and national governments started to panic, scientists began working on vaccines. Synthetic genes led to more than 200 different vaccines starting to be developed.[2] One firm, Moderna, managed to produce the first vials of its vaccine on 7 February, 31 days after the sequence of the virus was initially released. It took the company, in partnership with the National Institutes of Health, a mere six days to finalise the sequence of the vaccine, and a further 25 days to manufacture it.[3] Within 12 months of the virus being identified within humans, many countries had approved seven different effective vaccines.[4] Globally, nearly 24 million people had received a vaccine shot by the first anniversary of the virus sequence being made public.

The speed with which scientists went from identifying a novel pathogen to getting a working vaccine was unprecedented in human history. The meningitis vaccine took 90 years to develop, polio 45 years, measles a decade. Scientists created an effective vaccine for mumps in just four years: that was a proud achievement at the time. Given the virulence of SARS-CoV-2 and the effects it was having, no one could afford to wait that long. And so they didn't.

The medical wonder of the worldwide response was, of course, only one side of the story. The vaccine was developed so quickly because coronavirus had brought about a global crisis. The disease spread rapidly thanks to the highly networked economy of the Exponential Age. On 20 March, there were 10,000 Covid-19 deaths reported in total; on 1 April, the worldwide death toll had risen above 50,000; by September, it had exceeded 1 million. Within a year, there were 2.7 million deaths from Covid-19 reported globally, with the real toll estimated to be much higher. And the virus brought untold social turmoil. By the beginning of April 2020, the majority of countries around the world had gone into some form of lockdown.[5] The pandemic was the biggest economic event in modern times, on par with the Wall Street Crash of 1929 – leading to market turmoil, global scuffles over logistics and supply chains, and a fundamental shock to working practices as millions found themselves working remotely or, worse, out of work.

But the exponential speed of the spread of Covid-19 shouldn't overshadow the exponential speed of our response. The triumph of the vaccine reflected the transformative power of recent technologies, at every stage of the process. It started with genome sequencers.

By 2020, there were about 20,000 sequencers in labs, hospitals and universities around the world. They allow scientists to isolate a bit of DNA or RNA and transform it into a gene sequence – the string of letters that represent its component amino acids. This information is uploaded to the kinds of communal databanks discussed in the previous chapter – each a kind of Wikipedia for gene sequences, a space in which people can freely access and make use of genetic information. This technology underpinned scientists' lightning-fast analysis of Covid-19.

Moderna developed its vaccine with a 'modified messenger RNA', or mRNA, a method that involves teaching our cells how to create proteins that resist a virus. And it was able to design such a vaccine thanks to the power of machine learning – Moderna's platforms crunched millions of numbers to identify what concoction would spark the right response from human cells. It is a characteristically exponential process: according to Marcello Damiani, one of the firm's bosses, 'When we run experiments, we collect even more data. This allows us to build better algorithms, which helps build the next generation of medication. It's a virtuous cycle.'[6] The computational tools used by Moderna compressed some tasks that usually take years down to just hours.[7]

When it came to testing, digital platforms enabled the recruitment of large numbers of trial candidates on whom to test the vaccine. Without those electronic systems, recruiting enough volunteers might have taken months, if not years. During the roll-out, health services relied on databases, online booking systems and text messages to speed up vaccination rates. Databases kept track of who received the jab and who still needed it. Without this technology, many more would have died.

At the same time, exponential technologies defined our cultural and social responses to the pandemic. From the outset, scientists and governments were forced to wage war on online misinformation. Rumours, misinterpretations, deliberate falsehoods, anti-vaccination rhetoric, fake news – all cascaded across social networks. Researchers tracking the spread of ideas over different social networks tried to estimate how quickly Covid misinformation spread. It was highly contagious. The R0 number measures the transmissibility of a pathogen. If the R0 is above 1, a pathogen will spread exponentially; if the R0 is below 1, the spread will stop. The R0 for SARS-CoV-2 is estimated at much more than 3, the common cold between 2 and 3, and the seasonal flu around 1. When the researchers calculated the R0 of Covid misinformation on different social networks, they discovered that it ranged from 1.46 (for Reddit) to 2.24 (for Instagram). All driven by exponential technology – with a single click we could share our unsubstantiated thoughts with hundreds or thousands of people.

During lockdowns, billions of us were compulsorily billeted in our homes. For those with internet access and a computer, tablet or phone, creature comforts – from Netflix to online games – were not in short supply. We filled our fridges with groceries bought online, with DoorDash or Deliveroo providing a wide variety of choice if we wanted a treat. Companies and students flipped to remote work and learning, relying on videoconferencing. The Californian company Zoom became the standard bearer, its mosaic of just-too-small faces and exaggerated waves a familiar backdrop for many a working day.

If an earlier pandemic – say, SARS in 2003 – had struck as virulently as the Covid pandemic, life would have been different. Many of the services that were a lifeline in 2020 were less than a decade old – yet under lockdown, few could imagine a time before them. In 2010, computers were less powerful, phones dumber, broadband worse, social networks largely non-existent. YouTube was founded in 2005; Netflix launched online streaming in 2007; Zoom launched in 2011; DoorDash in 2013. Fewer than one in five US homes had broadband in 2003, and average internet speeds were a twentieth of those in 2020. Remote working as we know it would have been a pipe dream, because the infrastructure that enabled it in 2020 simply didn't exist then.

The way Covid-19 changed the world was, in short, a symptom of exponential technology. As the virus spread, it was both driven and combated by the innovations described in this book. It proclaimed, more loudly than any previous event, we are in the Exponential Age.

One of the themes of this book has been the way exponential technologies give us more from less. With every given year, technologies deliver more computational power and cheaper energy. They manipulate matter and engineer biology. They achieve more impressive things with less stuff and for less money. They let us develop vaccines, roll out e-commerce, pivot to online work. And it is a cyclical process. As demand for services grows, we march inexorably down the path of Wright's Law. Increased production drives greater efficiencies, ever-more transformative technologies, and increasingly connected networks of trade and information.

There are few signs of this shift slowing down. As new technologies – from AI to solar power – continue to improve, the Exponential Age will only accelerate. Scientific research will continue, and entrepreneurs will continue to use it to build revolutionary products. The exponential curve of technological performance – and social transformation – will continue its upward trajectory.

Yet for all its talk of our changing economy, politics and society, this book has rarely touched upon where the Exponential Age is taking us in the long run – over the course of decades, rather than years. And with good reason. Making predictions in the Exponential Age is a perilous business. Events have a way of spiralling off in unexpected directions.

We can, however, make out the direction of travel. Nobody knows how exactly exponential technologies will shape our lives. But some changes seem more likely than others.

A decade from the time of writing, computer power will be 100 times cheaper than it is now. A dollar spent will buy you at least 100 times more computing power than you get today. Understanding the implications of this is not easy. It's like asking someone in the early days of electricity, say in 1920, how they would use the amount of power those of us in the rich world today consume daily without thinking. Back then, the average British home consumed less than 50 watt-hours of electricity per day.[8] One hundred years later, the typical UK home guzzles 200 times as much electricity. Americans enjoy still more.

In 1920, it would have been hard to imagine the comforts of today: electric underfloor heating, LED lighting of any hue, tiny motors that open and close our blinds, toothbrushes that vibrate hundreds of times a second, vacuum cleaners that can suck up barely visible animal dander. These machines depend on vast amounts of electricity and a cascade of technologies layered on top of each other. So what might we do with a tremendous growth in computing power – at least 100-fold in any given decade and likely much more? It will create a similar discontinuity to the rapid growth of electricity, the effects of which are hard to predict.

A decade from now, renewable electricity from solar or wind plants will be five times cheaper, and perhaps more – likely a quarter of the 2020 price of natural gas. The batteries used in electric vehicles will probably be a third of the price they were in 2020. Renewable energy will have become the dominant form of electricity in major economies. It seems unlikely that any new sources of fossil fuel power will be coming online in advanced economies. Even in poorer countries, old gas and coal plants are being wound down. Oil companies may well have found better things to do with their time than pump oil. Abu Dhabi's investments in solar farms across the UAE and in Africa demonstrate the scale of this shift. But solar and wind power will only be part of the transition to a sustainable future: the methods of synthetic biology will allow us to shift away from using fossil fuels in pharmaceuticals and oil-derived materials such as plastics.

A decade from now, human genome sequences will be approaching a dollar apiece. There will be little reason not to sequence every person as part of their normal medical procedures. Building on the advances presaged by services like Ping An Good Doctor, you might have access to hospital-grade diagnostics from your neighbourhood pharmacy. Should you develop a rare illness, personalised medications could be your saviour. Cheap wearable devices and regular testing of your blood and microbiota might allow doctors to catch many conditions before they become serious. The combination of large-scale population data and personal information could help fine-tune diets for health.

A decade from now, scores of nanosatellites providing real-time sensor data might help us keep track of our natural resources – including coral reefs, ocean ecosystems and forests. Scientists plan to map and track every single one of the 3 trillion trees on the planet, to better understand rates of deforestation – which will be critical in managing our carbon budget.[9] Vertical farms could reduce our dependence on water and other resources, all the while providing healthier, more nutritious, food to our cities.

In short, we are entering an age of abundance. The first period in human history in which energy, food, computation and much else will be trivially cheap to produce. We could fulfil the current needs of humanity many times over, at ever-declining financial cost.

Yet this is only half the picture. The technology offers up new possibilities. But technology alone will not solve our problems.

We should be concerned with three issues. The first is that an abundance of stuff – energy, materials, healthcare, whatever – doesn't mean an absence of waste. Sure, some countries have decarbonised their electricity supply and broken the link between carbon output and GDP. But that doesn't eliminate resource use altogether. Solar panels and chips need sand; batteries need metals that must be yanked out of the ground; urbanisation will require buildings, roads, sewers and more. In fact, history has shown that the cheaper something becomes, the more people can afford it, and so the more demand there is. Supply catalyses demand: my new shelving, bought to help me tidy the piles of books on my desk, resulted in books both on straining shelves and in unruly piles on my desk. Exponential technologies by no means preclude excessive resource extraction.

Imagine that the resources we need to sequence a single human genome reduce by a factor of 100, pushing the cost of sequencing down much further. If, as a result of this lower cost, we end up sequencing a million times more genomes, our overall use of resources will have increased. The risk is that reduced costs lead to gross increases in consumption.[10] Exponential technologies can help us tackle human crises like climate change. But the technology alone is not enough. Without the right governance, they may still lead us down the road of overconsumption and environmental catastrophe.

The second issue is that technology is destabilising. At these exponential rates of improvement and deployment, technological innovation challenges well-established systems. As we've seen time and again in this book, technology has an earth-shattering effect on our social, economic and political institutions. New companies force out old ones. The relationship between workers and employers is transformed. Economic cooperation gives way to localised production. And the effects of this destabilisation are usually borne by those who can least afford them. Smaller, more vulnerable and low-tech companies are the ones that go bust. Workers with lower levels of training find themselves thrown into exploitative gig work. The localisation of supply chains is ruinous for the economies of the developing world. The negative effects of the Exponential Age are never evenly distributed.

A third concern is the dramatic shift in the locus of power that exponential technologies bring about. These technologies have contributed to the creation of superstar companies. And these firms have capabilities that even governments need to rely on. Such companies increasingly dominate aspects of society we have never thought of as the province of firms. But it is not just companies that are becoming more powerful. Unaccountable influence goes to the individuals and places – high-tech knowledge-workers, cyber criminals, malevolent states – that first understood the power of exponential technology.

The notion of the exponential gap is one attempt to synthesise each of these forces. All of them hint at a world changing faster than our systems and assumptions can handle. And any of these forces might lead us to dystopia. An era of constant military conflict; of increasingly powerful firms and powerless workers; of a hyper-commodified public sphere attuned to the needs of the people building the technologies and few others – all set against the backdrop of environmental crisis.

But this dystopian world is not inevitable. It is easy, when faced with transformative technologies, to become deterministic. We sometimes think that it is the force of technology alone that drives social change. Or assume that technology has its own path over which we have no control. Sheila Jasanoff, an ethicist based at Harvard University, critiques a widespread notion that technology 'possesses an unstoppable momentum, reshaping society to fit its demands'.[11] All too many people, Jasanoff says, think that to stand in the face of technology is to be somehow primitive, a Luddite. Technology, their argument goes, is not something that can be stopped.

Yet technology doesn't determine how it develops. We do.

This insight is key if we're to close the exponential gap. Throughout this book, I've identified the specific ways we might put technology back in the service of society. But more broadly, the solution to the exponential gap is a shift in mindset – one that acknowledges that we have agency over where technology will take us.

It is a shift with two main stages. The first is to acknowledge that technology's shape, direction and impact are not preordained. Of course, technology builds on what has come before – new innovations layer and combine with those of earlier generations. But its path is not set. We are the ones who decide what we want from the tools we build.

This is shown most clearly in the divergent ways technologies are used by different societies. Consider DDT, which was very helpful at killing mosquitos and slowing the spread of malaria, but also polluted food chains around the world. DDT was banned in the United States and United Kingdom. Yet for decades it was allowed in India, where preventing malaria was a priority – and so ecological damage was deemed an acceptable downside. Gig work tells a similar tale: in the UK, a rich country with a well-established formal labour market, gig-working platforms may be seen to unacceptably undermine workers' rights and protections. In India or Nigeria, the same technologies may bring benefits, as these countries have a more informal work market based on day labour. In other words, it is our choices and our circumstances that determine how technology is actually used. As the historian Melvin Kranzberg put it: 'The point is that the same technology can answer questions differently, depending on the context into which it is introduced and the problem it is designed to solve.'

But acknowledging that the direction of technology is not preordained doesn't amount to saying that technology isn't transformative. Elsewhere, Kranzberg writes that technology 'is neither good, nor bad, nor is it neutral'.[12] Like it or not, technology brings change. And so the second stage in our shift in mindset involves acceptance: recognising that while we can shape it, technology will nonetheless bring rapid, often unexpected dislocations. In the face of these changes, we must avoid the temptation to prevent experimentation, or choke the creative vigour of the market. Chaotic, unpredictable developments can be for the good – it's our responsibility to channel them when we can and manage their surprises when we can't.

As the Exponential Age accelerates, general purpose technologies will affect all of our most cherished institutions. Our societies are bound together by thousands of unspoken rules, norms, values and expectations. But technology disrupts them. As Jasanoff puts it, 'In all of its guises, actual or aspirational, technology functions as an instrument of governance.' It moulds all of the ethical, legal and social systems on which we depend.[13] We cannot prevent this moulding happening, but we can direct it.

Of course, changing how we think about the pace of change won't, on its own, close the exponential gap. We need policies, social movements and new forms of political and economic organisation too. And developing these won't be easy, given just how many domains of our politics, economy and society are impacted by exponential technology.

As this book has shown, the solutions are varied. Our economy is increasingly dominated by enormous, sometimes oligopolistic companies – forged by increasing returns to scale. These firms operate the infrastructure that underpins the modern world: from the technical internet services that allow data to flow freely, to the software that helps our businesses to function. These firms defy our standard assumptions about the limits of a company's growth, and about what monopoly looks like. And so we need a new conception of what market dominance is – and to place new regulatory and societal expectations on the companies that start to look monopolistic.

In the workplace, our assumptions about the relationship between an employer and a worker are being eroded. We are moving towards a world dominated by platforms, which may leave workers in a precarious position. And these platforms will increasingly manage workers by algorithms optimised for output and efficiency. The shift in organisational structure and production methods can result in a declining worker share in profits. And so we need new ways to empower working people. Workers, whether formally employed or gig workers, need to be guaranteed dignity, flexibility and security – so they can continually adapt to the rapidly changing workplace, without their lives becoming unbearable. The best way to achieve this is through new forms of collective action, which will allow workers to exert pressure on their employers.

Our sense of place is being upturned as well. As supply chains are re-localised, regional economies in the developed world will become more independent. And cities will grow in stature too, drawing in the talent of a region and taking up the mental space of political leaders. Rural and small-town areas are at risk of getting left behind. Issues that we have thought of as national are becoming regional or even local – how will the structure of power within and between nations need to change to reflect that? On such matters, government needs to be closer to the people it serves. But we also need new organisations to maintain international cooperation, to prevent much of the world being shut out of the benefits of exponential technology.

This return to the local could lead to conflict between nations and between regions. And this risk is exacerbated by new technologies, which make attacks – from drone swarms to constant cyber warfare – much easier. We risk descending into a period of dramatic disorder. In response, nations need to strengthen their defences, and teach their populations to be more resilient to attacks. At the same time, we need new rules and norms of war – to better de-escalate conflicts, and to stop the constant proliferation of new weapons.

Finally, we need to rethink the relationship between citizen and society, particularly surrounding the role of the market in our lives. As ever-greater parts of our lives are turned into commodities – from our public conversations to our private data – we need to make sure that some areas fall beyond the remit of markets. We can do that by making the technology giants more transparent in their decisions, and by explicitly guaranteeing certain rights to digital citizens. At the same time, new systems of common ownership and control would allow our communities – rather than private companies, controlled by the few – to harness the power of exponential technology.

At first glance, these policies might seem disparate. Yet there are a few threads that hold them together. It is these unifying themes that anyone serious about tackling the exponential gap should keep in mind. When water turns to steam, we can still make use of its power. But we need new tools to handle it – or we risk getting scalded. And so we need a new set of tools to manage our phase transition into the Exponential Age.

The first principle is commonality. As the world changes at an increasing pace, no state, business or worker will be able to keep up alone. And so we need to build institutions that allow disparate groups of people to work together, to cooperate and to exchange ideas. This principle underpins many of the policies identified in this book. Interoperability, for example, is about encouraging different businesses to work together. It is about ensuring that as consumers (and customers) we can easily use services from different suppliers, without being locked in to any one of them. Cooperation between nations with similar interests, in the form of new intergovernmental bodies, will help reduce the risk of global conflict. Cooperation will also drive the new wave of commonly owned and commonly run organisations – commons approaches to data, and even to creating services, for example. And a greater emphasis on worker collectives will allow everyone to get a better bargain in the face of volatile shifts in the world of work.

The second principle is resilience. The world is developing faster than ever, and we need institutions that are sturdy enough to handle constant change. When we talk about new forms of welfare – the Danish 'flexicurity' model identified in Chapter 5, for example – we are trying to build systems that don't collapse under the strain of a rapidly changing labour market. And when we emphasise the need for new digital rights for citizens, we are trying to create a bedrock of basic protections that will remain in place whatever direction the digital platforms develop in. These are only a couple of suggestions. All organisations will need to consider how to develop sturdy systems – designed with resilience in mind rather than with resilience as an afterthought.

Resilience, though, does not mean rigidity. And so our third principle is flexibility. Our institutions need to be able to adapt quickly, as society around them changes. Whether it is ignoring cyber threats, or failing to update employment legislation, or being slow to rethink the nature of monopoly, our systems have been lackadaisical in their response to the changing technological order. Many of the suggestions in this book have been about trying to help institutions adapt more quickly. When we hand down power to the local level, we create political units that are small enough to be agile. When we increase the transparency of big tech, we become more likely to spot incipient problems and respond rapidly – before anything gets out of control. These are just a few examples. In the Exponential Age, all institutions need to be serious about flexibility. If they are too rigid, they will be outpaced by our changing world.

Provided you are reading this book at some point before 2040, chances are you have lived more of your life before the Exponential Age than during it. And that means that the institutions around you were, for most of your life, adequate. The industrial age brought new technologies: the internal combustion engine, the telephone, electricity. In response, humans built social institutions that adapted these technologies to their needs: secure jobs and collective bargaining for workers; national electricity boards; road safety manuals. As the twentieth century rolled on, these institutions gradually morphed in line with the changing needs of society. The working norms of the 1990s differed from those of the 1930s. And this gradual pace of change was fine. Technologies were not developing at disorientating rates. There were massive shocks – think of World War Two and the 1973 oil crisis – but by and large technology delivered a period of evolution, not disjuncture.

But we are now well into the Exponential Age. Destabilisation and disorder are constant. And so we need a new social settlement – one that is fit for an era of constant change. The principles that underpin this book – commonality, resilience and flexibility – may seem distant, but they are the only way I can see to make our institutions function in the Exponential Age. And, if they seem fanciful, remember that we have transformed our institutions before. The ideas of universal suffrage, permanent employee contracts and global supply chains once seemed the realm of fiction too.

That humans have created new systems before should give us cause for hope. In the Exponential Age, technology is unpredictable. It is hard to say how new innovations will transform our society, as they constantly interact with our approaches to business, work, place, conflict and politics. But after all the technological revolutions of the past, humans have found ways to thrive. Modern history is defined by two great forces: the extraordinary power of technological change; and humans' ingenuity in forging the world we want in response. Technology, it turns out, is something that we can control.

And because it is in our control, regardless of how sophisticated it may become, technology always has the potential to be a force for good. Water turns to steam, but that doesn't stop us harnessing its power.

Acknowledgements

A book is both the work of an author and the work of a collective. The writer travels alone, yet their solo journey is not possible without a network of contributors and supporters.

To my research team, led by Marija Gavrilov and assisted by Sanjana Varghese, Emily Judson and Joseph Dana. You were bombarded with questions on everything from labour relations in turn-of-the-century New York to the likely timeline for quantum supremacy; from theories of institutional change to the limits of photolithography in semiconductor manufacturing. You not only survived, but flourished.

My editor, Rowan Borchers at Penguin Random House, was an instigator and supporter of the project, helping a first-time author figure out how to tell his story. Rowan wrestled the manuscript into shape, gently delivering the most savage cuts. Gemma Wain's copy editing was meticulous and sensitive. Isabelle Ralphs, Anna Hibberd and Adrienne Fontaine ensured word got out. My agent Jeff Shreve proved an invaluable sounding board as the project took shape. In the US, Scott Waxman, Keith Wallman and Diversion Books offered a dynamic partnership.

I'm indebted to the many readers who tirelessly read early drafts. The few brave souls who read preliminary iterations of the entire book: Mark Bunting, Kevin Werbach and Tom Glocer. And many others who provided invaluable comments that helped shape key chapters: Laure-Claire Reillier, Libby Kinsey, Tom Wheeler, Carly Kind, Sameh Wahba, Giedrimas Jeglinskas, Christina Colclough, Paul Daugherty, Matthew

Taylor, Stian Westlake, Robbie Stamp, Tim O'Reilly, Pascal Finette and Dan Gillmor.

The special projects and podcast teams at *Exponential View*, including Fred Casella, Ilan Goodman, Katie Irani, Elise Thomas, Diana Fox Carney, Joanna Jones, Jayne Packer, Nasos Papadopoulos and Bojan Sabioncello, kept the guests – and the insights – flowing.

This book would also not have been possible without the readers of my newsletter, *Exponential View*, and in particular those who gave it the early momentum – including Laurent Haug, John Henderson, Martin De Kujper, Fred Wilson, Hamish Mackenzie, Mustafa Suleyman, Kenn Cukier and Daniel Ek. A special mention goes out to members of the Exponential Do community, whose vigorous and ongoing discussions helped illuminate many issues. Particular thanks to Zavain Dar and Tuan Pham, who enticed me into presenting the earliest cut of these ideas in December 2016; and to Shamil Chandaria, who first suggested that my thesis not only could make a book but should make a book. Plus a grateful call-out to all the founders and clients whose calls I didn't return during crunch periods of research and writing.

The work of three academics was key in helping me understand the nature of technology and its place in society: Carlota Perez's work on technological revolutions was instrumental in helping me think about the interrelationship between technology and economic paradigms; W. Brian Arthur's research on increasing returns to scale and the ecological nature of technology also provided a key foundation. Vaclav Smil's work on energy and its impact on human and societal development was also helpful.

Beyond this, many others have helped, perhaps unknowingly, through their discussions with me, including Rumman Chowdhury, Bill Janeway, Anders Wijkman, Marko Ahtisaari, Toomas Hendrik Ilves, David Kausman, Alexandra Mousavizadeh, Mark Evans, Martin Tisné, Stephanie Hare, Reid Hoffman, Tom Loosemore, Celine Herweijer, Dan Elrond, Nicolaus Hencke, Ivan Ostojic, Diana Foltean, Ray Eitel Porter, Barney Pell, Elisabeth Ling, Raj Jena, Manar Hussain, Christopher Mims,

Benedict Evans, Nick Russell, Nat Bullard, Matthew Stoller, John Battelle, Velimir Gašparović, Farhan Lalji, James Wang, Salim Ismail, Yuri van Geest, Salman Malik, Tom Kelley, Simon Daniel, Gerd Leonhard, Rob McCargow, David Galbraith, Tabitha Goldstaub, David Giampaolo, Ed Vaizey, Paul Nemitz, Wendy Hall, Tim Gardam, Mayra Valderama, Adrian Weller, Eleanor O'Keeffe, Tom Standage, Jerry Li, Reema Patel, Jeni Tennison, Nigel Shadbolt, Blair Sheppard, Horace Dediu, Ramez Naam, Diane Coyle, Nicolas Colin, Christian Printzell Halvorsen, Terje Seljeseth and Brett Frischmann. Many investors, over the decades, helped me understand the market, including Leila Zenga, Chrys Chrysanthou, Russell Buckley, Liz Broderick, Albert Wenger, Saul Klein, Reshma Sohoni, Hussein Kanji, Sean Park, Ciaran O'Leary, Jason Whitmire, Christian Hernandez, Eileen Tso, Yann Ranchere, Ash Fontana and Jim Pallotta.

Several people taught me how to write and think more clearly, including Frances Cairncross, Bill Emmott, Alan Rusbridger, John Micklethwait, John Peet, Nick Passmore, Geoff Carr, Gideon Lichfield, Mark Roberts, Oliver Morton and Ditlev Schwanenflugel. Further thanks to the many teachers who helped me understand the world of computing, politics and economics, including John Wilcox, Stephen Bishop, Nick Lord, Ray Bradley, Mike Clugston, Don Markwell, Vijay Joshi, the late John Lucas, the late David Bostock and the late Jack Schofield.

Many thanks to the dozens of guests on my podcast whose ideas have helped enrich my thesis, including Laetitia Vitaud, Bill Janeway, Carissa Véliz, Tony Blair, Demis Hassabis, Sam Altman, Philip Auerswald, Scott Santens, Jeff Sachs, Andrew Yang, Jack Clark, Trent McConaghy, Michael Liebreich, Casper Klynge, Kate Raworth, Sir Richard Barrons, Joanna Bryson, Stuart Russell, Cory Doctorow, Kai-Fu Lee, Matt Clifford, Marietje Schaake, Yuval Noah Harari, Mariana Mazzucato, Mike Zelkind, Josh Hoffman, Binyamin Applebaum, Kate Crawford, Matt Ocko, Jeremy O'Brien, Sam Altman, Audrey Tang, Vijay Pande, Matt Clifford, Fei-Fei Li, Adena Friedman, Kersti Kaljulaid, Astro Teller, Deep Nishar, Cesar Hidalgo, Ian Bremmer, Brad Smith, Nicole Eagan, Meredith

Whittaker, Gary Marcus, Andrew Ng, Shoshana Zuboff, Jürgen Schmidhuber, Gina Neff, Missy Cummings, Eric Topol, Cathie Wood, Michael Liebreich, Mariarosaria Taddeo and Ronit Ghose.

Thanks to my parents, Aleem and Kaneez Azhar, who introduced me to the ideas of economics and its development impact, especially in those early years. And an extra mention for my mum, who was responsible for getting computers into our home in the early 1980s. My sister Lubna has always been an encouraging spirit. My parents-in-law, Hatim and Asma Suterwalla, were a source of regular support, and my sister-in-law, Mumtaz Suterwalla, a constant positive presence.

My children, Salman, Sophie and Jasmine, have been on this exponential journey with me for years, surprising me with their humour and insight.

Finally, to Shehnaz Suterwalla, who invested more than her fair share of time hearing the argument, challenging it, and helping me sharpen it when it was nebulous and blancmange-like – all while moving her own projects, and our family, forward.

Notes

PREFACE: THE GREAT TRANSITION

1 Stefan Thurner, Peter Klimek and Rudolf Hanel, *Introduction to the Theory of Complex Systems* (Oxford: Oxford University Press, 2018), p. 8 <https://www.oxfordscholarship.com/view/10.1093/oso/9780198821939.001.0001/oso-9780198821939> [accessed 19 July 2020].

2 See also Ricard V. Solé, *Phase Transitions*, Primers in Complex Systems (Princeton, NJ: Princeton University Press, 2011).

3 See for example: 'Trouble with a Telephone', *New York Times*, 25 September 1898; '"Hobbles" and Dancing; Women's New Fashions Will Influence Length of Steps', *New York Times*, 13 November 1910; 'Judge rails at jazz and dance madness', *New York Times*, 14 April 1926.

4 *Edelman Trust Barometer 2020* (Edelman, 19 January 2020), p. 15 <https://www.edelman.com/sites/g/files/aatuss191/files/2019-02/2019_Edelman_Trust_Barometer_Global_Report.pdf> [accessed 16 June 2021].

5 Solé, *Phase Transitions*, pp. 2–11, pp. 132–136, pp. 186–187

6 Eric Schmidt and Jared Cohen, *The New Digital Age: Reshaping the Future of People, Nations and Business* (New York: Alfred A. Knopf, 2013), p. 55.

7 Peter H. Diamandis and Steven Kotler, *Abundance: The Future Is Better Than You Think* (New York: Free Press, 2014), p. 213.

8 See, for example, Chris Stokel-Walker, 'Politicians Still Don't Understand the Tech Companies They're Supposed to Keep in Check. That's a Problem', *Business Insider*, 10 October 2020 <https://www.businessinsider.com/tiktok-ban-hearings-politicians-senators-know-nothing-about-tech-2020-10> [accessed 12 April 2021].

9 Charles Percy Snow, *The Two Cultures* (Cambridge University Press, 2012).

CHAPTER 1: THE HARBINGER

1 In other words, it was a Turing Machine – so named after British mathematician Alan Turing, who devised much of the theory behind computer science. Turing's tragic death in 1954 meant he never had access to a computer as generally capable as the ZX81, with its 1,024 bytes of memory storage capable of crunching through a superhuman half a million instructions per second.

2 G. E. Moore, 'Cramming More Components onto Integrated Circuits', *Proceedings of the IEEE*, 86(1), 1965, pp. 82–85 <https://doi.org/10.1109/JPROC.1998.658762>.

3 Newton's laws work at the scale of the everyday and in what is known as 'inertial reference frames'. At the very small – the levels of atoms and smaller – we need to rely on quantum physics to describe what is going on. 'Non-inertial reference frames', such as those found when studying cosmology, require different approaches.

4 Cyrus C. M. Mody, *The Long Arm of Moore's Law: Microelectronics and American Science*, Inside Technology (Cambridge, MA: The MIT Press, 2017), pp. 5 and 125.

5 Computer History Museum, '1958: Silicon Mesa Transistors Enter Commercial Production'.

6 Ray Kurzweil, 'Average Transistor Price', Singularity.com <http://www.singularity.com/charts/page59.html> [accessed 10 March 2021].

7 Dan Hutcheson, 'Graphic: Transistor Production Has Reached Astronomical Scales', *IEEE Spectrum*, 2 April 2015 <https://spectrum.ieee.org/computing/hardware/transistor-production-has-reached-astronomical-scales> [accessed 4 December 2020].

8 James W. Cortada, *The Computer in the United States: From Laboratory to Market* (Armonk, NY: M. E. Sharpe, 1993), p. 117; 95 per cent of these were in the US. See also 'Early Popular Computers, 1950 – 1970', Engineering and Technology History Wiki <https://ethw.org/Early_Popular_Computers,_1950_-_1970> [accessed 28 July 2020].

9 Horace Dediu, personal correspondence with the author, 1 December 2016.

10 In 1940, 65 per cent of American homes had flush toilets. By 1960, it was 90 per cent. 'Historical Census of Housing Tables – Sewage Disposal' <https://www.census.gov/data/tables/time-series/dec/coh-sewage.html> [accessed 31 July 2020].

11 Paul R. Liegey, 'Microwave Oven Regression Model', U.S. Bureau of Labor Statistics <https://www.bls.gov/cpi/quality-adjustment/microwave-ovens.htm> [accessed 10 March 2021].

12 Alex Wilhelm, 'Charting Bird and Lime's Rapid Growth', *Crunchbase News*, 20 September 2018 <https://news.crunchbase.com/news/charting-bird-and-limes-rapid-growth/> [accessed 31 July 2020].

13 Computers are so cheap, they turn up in surprising places. Disposable pregnancy tests have a small computer in them – used once and thrown away.

14 Charlotte Burgess, 'Future Banking: Creating an "Incumbent Challenger"', *Finovate*, 2020 <https://finovate.com/future-banking-creating-an-incumbent-challenger/> [accessed 12 April 2021].

15 Author's analysis at Azeem Azhar, 'Big Ideas for 2021; China's Civil-Military Fusion; Tesla's Infrastructure Advantage; the Climate Decade; Elephant Body Fat & Hegelian Enjoyment ++ #307', *Exponential View*, <https://www.exponentialview.co/p/ev-307> [accessed 11 March 2021].

16 David Rotman, 'We're Not Prepared for the End of Moore's Law', *MIT Technology Review*, 24 February 2020 <https://www.technologyreview.com/2020/02/24/905789/were-not-prepared-for-the-end-of-moores-law/> [accessed 11 March 2021].

17 Ray Kurzweil, 'The Law of Accelerating Returns', Kurzweil.net, 2001 <https://www.kurzweilai.net/the-law-of-accelerating-returns> [accessed 29 July 2020].

18 Kurzweil also predicts a 'singularity' where the capacities of computers exceed those of the human brain. This argument is not relevant to my argument, so I don't consider it here.

19 Azeem Azhar, 'Beneficial Artificial Intelligence: My Conversation with Stuart Russell', *Exponential View*, 22 August 2019 <https://www.exponentialview.co/p/-beneficial-artificial-intelligence> [accessed 16 April 2021].

20 Dario Amodei and Danny Hernandez, 'AI and Compute', OpenAI, 16 May 2018 <https://openai.com/blog/ai-and-compute/> [accessed 12 January 2021].

21 Charles E. Leiserson et al., 'There's Plenty of Room at the Top: What Will Drive Computer Performance after Moore's Law?', *Science* 368(6495), June 2020 <https://doi.org/10.1126/science.aam9744>.

22 Jean-François Bobier et al., 'A Quantum Advantage in Fighting Climate Change', BCG Global, 22 January 2020 <https://www.bcg.com/publications/2020/quantum-advantage-fighting-climate-change> [accessed 23 March 2021].

CHAPTER 2: THE EXPONENTIAL AGE

1 Bryno85, 'James Bond Movies Ranked via IMDb Rating', IMDb, 18 October 2015 <http://www.imdb.com/list/ls078535153/> [accessed 2 August 2020].

2 Gregory F. Nemet, *How Solar Energy Became Cheap: A Model for Low-Carbon Innovation* (London; New York, NY: Routledge/Taylor & Francis Group, 2019), p. 62.

3 Nathaniel Bullard, 'The Energy Revolution That Started in 1954 Is Reaching Its Crescendo', *Bloomberg Green*, 23 April 2020 <https://www.bloomberg.com/news/articles/2020-04-23/the-energy-revolution-that-started-in-1954-is-reaching-its-crescendo> [accessed 3 August 2020].

4 Ramez Naam, 'The Exponential March of Solar Energy', *Exponential View*, 14 May 2020 <https://www.exponentialview.co/p/-the-exponential-march-of-solar-energy> [accessed 1 August 2020]."

5 'Levelized Cost of Energy and Levelized Cost of Storage – 2020', Lazard, 19 October 2020 <http://www.lazard.com/perspective/levelized-cost-of-energy-and-levelized-cost-of-storage-2020/> [accessed 4 December 2020].

6 Sheldon Reback, 'Solar, Wind Provide Cheapest Power for Two-Thirds of Globe', *Bloomberg*, 27 August 2019 <https://www.bloomberg.com/news/articles/2019-08-27/solar-wind-provide-cheapest-power-for-two-thirds-of-globe-map> [accessed 16 April 2021].

7 Justin Rowlatt, 'What the Heroin Industry Can Teach Us about Solar Power', *BBC News*, 27 July 2020 <https://www.bbc.com/news/science-environment-53450688> [accessed 3 August 2020].

8 Jon Moore and Nat Bullard, *BNEF Executive Factbook 2021* (BloombergNEF, 2 March 2021).

9 Kris A. Wetterstrand, 'DNA Sequencing Costs: Data from the NHGRI Genome Sequencing Program', National Human Genome Research Institute <https://www.genome.gov/about-genomics/fact-sheets/DNA-Sequencing-Costs-Data> [accessed 15 February 2020].

10 Antonio Relgado, 'China's BGI Says It Can Sequence a Genome for Just $100', *MIT Technology Review*, 26 February 2020 <https://www.technologyreview.com/s/615289/china-bgi-100-dollar-genome/> [accessed 3 March 2020].

11 Antonio Relgado, 'EmTech: Illumina Says 228,000 Human Genomes Will Be Sequenced This Year', *MIT Technology Review*, 24 September 2014 <https://www.technologyreview.com/2014/09/24/111298/emtech-illumina-

says-228000-human-genomes-will-be-sequenced-this-year/> [accessed 28 July 2020].”

12 Zachary D. Stephens et al., ‘Big Data: Astronomical or Genomical?’, *PLoS Biology*, 13(7), July 2015 <https://doi.org/10.1371/journal.pbio.1002195>.

13 See discussion at ‘Why Has The Cost Of Genome Sequencing Declined So Rapidly?’ Biostars forum <https://www.biostars.org/p/42753/> [accessed 28 July 2020] and at ‘Next-Generation-Sequencing.v1.5.4 @albertvilella’ <https://docs.google.com/spreadsheets/u/0/d/1GMMfhyLK0-q8XkIo3YxlWaZA5vVMuhU1kg41g4xLkXc/edit#gid=1569422585> [accessed 28 July 2020].

14 Michael Chui and others, ‘The Bio Revolution: Innovations Transforming Economies, Societies and Our Lives.’ (McKinsey Global Institute, 2020)

15 Conversations with the author, November 2018 and August 2020. See also India Block, ‘World’s Largest 3D-Printed Building Completes in Dubai’, *Dezeen*, 22 December 2019 <https://www.dezeen.com/2019/12/22/apis-cor-worlds-largest-3d-printed-building-dubai/> [accessed 1 August 2020].

16 Christopher L. Benson, Giorgio Triulzi and Christopher L. Magee, ‘Is There a Moore’s Law for 3D Printing?’, *3D Printing and Additive Manufacturing*, 5(1), March 2018, pp. 53–62 <https://doi.org/10.1089/3dp.2017.0041>.

17 Wohlers Associates, *Wohlers Report 2020*.

18 Richard Lipsey, Kenneth Carlaw and Clifford Bekar, *Economic Transformations: General Purpose Technologies and Long-Term Economic Growth* (Oxford University Press, 2006).

19 Lipsey et al., *Economic Transformations*, p. xvi.

20 James Bessen, *Learning by Doing: The Real Connection between Innovation, Wages, and Wealth* (New Haven: Yale University Press, 2015), p. 49.

21 Carlota Perez, *Technological Revolutions and Financial Capital: The Dynamics of Bubbles and Golden Ages* (Cheltenham: Edward Elgar, 2003).

22 ‘Internet Growth Statistics 1995 to 2021 – the Global Village Online’ <https://www.internetworldstats.com/emarketing.htm>

23 Steven Tweedie, ‘The World’s First Smartphone, Simon, Was Created 15 Years before the iPhone’, *Business Insider*, 14 June 2015 <https://www.businessinsider.com/worlds-first-smartphone-simon-launched-before-iphone-2015-6> [accessed 21 February 2021].

24 Andrew Meola, ‘Rise of M-Commerce: Mobile Ecommerce Shopping Stats & Trends in 2021’, *Business Insider*, 30 December 2020 <https://

www.businessinsider.com/mobile-commerce-shopping-trends-stats> [accessed 21 February 2021].

25 T. P. Wright, 'Factors Affecting the Cost of Airplanes', *Journal of the Aeronautical Sciences*, 3(4), February 1936, pp. 122–128 <https://doi.org/10.2514/8.155>.

26 Béla Nagy et al., 'Statistical Basis for Predicting Technological Progress', *PLOS ONE*, 8(2), 2013, e52669 <https://doi.org/10.1371/journal.pone.0052669>.

27 Peter Ha, 'All-TIME 100 Gadgets', *Time*, 25 October 2010 <http://content.time.com/time/specials/packages/article/0,28804,2023689_2023703_2023613,00.html> [accessed 21 February 2021].

28 'The Drive to Decarbonize: Ramez Naam in Conversation with Azeem Azhar', *Exponential View with Azeem Azhar* [podcast], 15 April 2020 <https://hbr.org/podcast/2020/04/the-drive-to-decarbonize> [accessed 21 February 2021].

29 Marcelo Gustavo Molina and Pedro Enrique Mercado, 'Modelling and Control Design of Pitch-Controlled Variable Speed Wind Turbines', in Ibrahim H. Al-Bahadly, ed., *Wind Turbines* (InTech, 2011), p. 376 <https://doi.org/10.5772/15880>.

30 Vaclav Smil, 'Wind Turbines Just Keep Getting Bigger, But There's a Limit', *IEEE Spectrum*, 22 October 2019 <https://spectrum.ieee.org/energy/renewables/wind-turbines-just-keep-getting-bigger-but-theres-a-limit> [accessed 14 March 2021].

31 Hyejin Youn et al., 'Invention as a Combinatorial Process: Evidence from US Patents', *Journal of the Royal Society Interface*, 12(106), May 2015, 20150272 <https://doi.org/10.1098/rsif.2015.0272>.

32 Jonathan Postel, *Simple Mail Transfer Protocol* (Information Sciences Institute, August 1982) <https://tools.ietf.org/html/rfc821> [accessed 28 March 2021].

33 David Crocker, *Standard for the Format of ARPA Internet Text Messages* (University of Delaware, August 1982) <https://tools.ietf.org/html/rfc822> [accessed 28 March 2021].

34 Azeem Azhar, 'Disrupting the Insurance Industry with AI', *Exponential View with Azeem Azhar* [podcast], 14 August 2019 <https://hbr.org/podcast/2019/08/disrupting-the-insurance-industry-with-ai>.

35 Author's analysis of the UCS Satellite Database, 8 December 2005 <https://www.ucsusa.org/resources/satellite-database> [accessed 16 February 2020].

36 Tereza Pultrova, ‘ArianeGroup Futurist Sees Smallsat Standardization as Key for Timely Launch’, *SpaceNews*, 25 October 2017 <https://space-news.com/arianegroup-futurist-sees-smallsat-standardization-as-key-for-timely-launch/> [accessed 21 February 2021].

37 Ian J. Goodfellow et al., ‘Generative Adversarial Networks’, Cornell University Machine Learning Statistics, 10 June 2014 <https://arxiv.org/abs/1406.2661> [accessed 10 February 2020].

38 Until the Industrial Revolution, information travelled around the world at about 1 mph; this increased to around 3–4 mph in the 1800s. The arrival of the telegraph in the latter part of that century increased the speed of transmission by nearly 100 times. The internet has made the spread of information essentially instantaneous. See Gregory Clark, *A Farewell to Alms: A Brief Economic History of the World*, The Princeton Economic History of the Western World (Princeton, NJ: Princeton University Press, 2007), pp. 306–307.

39 ‘ArXiv Usage Statistics’, Cornell University ArXiv <https://arxiv.org/help/stats> [accessed 13 April 2021].

40 Martin Rittman, ‘Research Preprints: Server List’, <https://docs.google.com/spreadsheets/u/4/d/17RgfuQcGJHKSsSJwZZn0oiXAnimZu2sZsWp8Z6ZaYYo/edit#gid=0> [accessed 13 April 2020].

41 ‘COVID-19 Primer’ <https://covid19primer.com/> [accessed 30 November 2020].

42 ‘The 2020 State of the Octoverse’, Github Inc. <https://octoverse.github.com/> [accessed 6 December 2020].

43 Marc Levinson, *Outside the Box: How Globalization Changed from Moving Stuff to Spreading Ideas* (Princeton, NJ: Princeton University Press, 2020), pp. 61–67.

44 ‘HMM Algeciras: World’s Biggest Container Ship Arrives in Essex’, *BBC News*, 14 June 2020 <https://www.bbc.com/news/uk-england-essex-53041733> [accessed 7 December 2020].

45 ‘Container Port Traffic (TEU: 20 Foot Equivalent Units)’, World Bank Data, <https://data.worldbank.org/indicator/IS.SHP.GOOD.TU> [accessed 31 July 2020].

46 Binyamin Appelbaum, quoted in Azeem Azhar, ‘How Free-Market Economists Got It Wrong’, *Exponential View*, 27 December 2019 <https://www.exponentialview.co/p/-how-free-market-economists-got-it> [accessed 7 December 2020].

47 Milton Friedman, 'A Friedman Doctrine – The Social Responsibility of Business Is to Increase Its Profits', *New York Times*, 13 September 1970 <https://www.nytimes.com/1970/09/13/archives/a-friedman-doctrine-the-social-responsibility-of-business-is-to.html> [accessed 13 July 2020].

48 Dominic Ponsford, 'Economist Readership Analysed: Detailed Breakdown of Brand Reach for H1 2020', *Press Gazette*, 14 August 2020 <https://www.pressgazette.co.uk/economist-readership-brand-reach/> [accessed 7 December 2020].

CHAPTER 3: THE EXPONENTIAL GAP

1 The UK's national R&D spend was £37 billion, or approximately $47 billion. See 'Research and Development Expenditure', Office for National Statistics <https://www.ons.gov.uk/economy/governmentpublicsectorandtaxes/researchanddevelopmentexpenditure> [accessed 24 July 2020].

2 *Federal Research and Development (R&D) Funding: FY2020*, Congressional Research Service, 18 March 2020 <https://fas.org/sgp/crs/misc/R45715.pdf>.

3 Trefis Team, 'How Big Is Roche's R&D Expense?', *Forbes*, 19 December 2019 <https://www.forbes.com/sites/greatspeculations/2019/12/19/how-big-is-roches-rd-expense/> [accessed 24 July 2020].

4 Matthew Chapman, 'Are Retail Labs Really the Path to Innovation?' *Campaign Live*, 2 February 2016 <https://www.campaignlive.co.uk/article/retail-labs-really-path-innovation/1381099> [accessed 24 July 2020].

5 'Werner Vogels, Interview with Jon Erlichman of Bloomberg', 21 July 2020 <https://twitter.com/JonErlichman/status/1285628647609638915> [accessed 24 July 2020].

6 The other is people. People costs didn't deliver the same levels of deflation.

7 600 million litres is 600,000 cubic metres, or a cuboid 100 metres x 100 metres x 60 metres high.

8 Al Bartlett, 'Arithmetic, Population and Energy – a Talk by Al Bartlett' <https://www.albartlett.org/presentations/arithmetic_population_energy.html> [accessed 3 December 2020].

9 Joanna Stern, 'TikTok?! Clout-Chasing Millennial Learns About Memes and More', WSJ Video, 23 January 2020 <https://www.wsj.com/video/series/joanna-stern-personal-technology/tiktok-clout-chasing-millennial-learns-

about-memes-and-more/3C218B25-59AA-437C-BE7A-3F215B786DDA> [accessed 30 July 2020].

10 Of course, there are stories from antiquity that tackle this question as well. These normally tell the story of a vizier who asks to be rewarded by grains of rice placed on the squares of a chessboard. For every subsequent square, he asks for the amount of rice to be doubled. It is an exponential process that reaches astronomical quantities by the end of the board.

11 See, for example, Victor Stango and Jonathan Zinman, 'Exponential Growth Bias and Household Finance', *Journal of Finance*, 64(6), December 2009, pp. 2807–2849 <https://doi.org/10.1111/j.1540-6261.2009.01518.x> and Matthew R. Levy and Joshua Tasoff, 'Exponential-Growth Bias and Overconfidence', *Journal of Economic Psychology*, 58, 2017, pp. 1–14 <https://doi.org/10.1016/j.joep.2016.11.001>.

12 Johan Almenberg and Christer Gerdes, 'Exponential Growth Bias and Financial Literacy', *Applied Economics Letters*, 19(17), 2012, pp. 1693–1696 <https://doi.org/10.1080/13504851.2011.652772>.

13 William A. Wagenaar and Sabato D. Sagaria, 'Misperception of Exponential Growth', *Perception & Psychophysics*, 18(6), November 1975, pp. 416–422 <https://doi.org/10.3758/BF03204114>.

14 Fernand Braudel, *The Mediterranean and the Mediterranean World in the Age of Philip II* (New York: Harper & Row, 1972), p. 20.

15 Pascal Boyer and Michael Bang Petersen, 'Folk-Economic Beliefs: An Evolutionary Cognitive Model', *Behavioral and Brain Sciences*, 41, 2018, E158 <https://doi.org/10.1017/S0140525X17001960>.

16 Duff McDonald, *The Firm: The Story of McKinsey and Its Secret Influence on American Business* (New York: Simon & Schuster, 2014), pp. 178–179.

17 'Planet of the Phones', *The Economist*, 26 February 2015 <https://www.economist.com/leaders/2015/02/26/planet-of-the-phones> [accessed 15 March 2021].

18 Simon Evans, 'Solar Is Now "Cheapest Electricity in History", Confirms IEA', *Carbon Brief*, 13 October 2020 <https://www.carbonbrief.org/solar-is-now-cheapest-electricity-in-history-confirms-iea> [accessed 18 December 2020].

19 Ray Kurzweil, *The Age of Spiritual Machines: When Computers Exceed Human Intelligence* (New York, NY: Penguin, 2000).

20 Suzana Herculano-Houzel, 'The Human Brain in Numbers: A Linearly Scaled-up Primate Brain', *Frontiers in Human Neuroscience*, 3 November 2009 <https://doi.org/10.3389/neuro.09.031.2009>.

21 Carl Zimmer, '100 Trillion Connections: New Efforts Probe and Map the Brain's Detailed Architecture', *Scientific American*, January 2011 <https://doi.org/10.1038/scientificamerican0111-58>.

22 Even if we could build a machine with the complexity of the human brain – comprising artificial rather than real neurons, and connections between them – it isn't clear this would give rise to anything that can do what the human brain does.

23 Graham Rapier, 'Elon Musk Says Tesla Will Have 1 Million Robo-Taxis on the Road Next Year, and Some People Think the Claim Is So Unrealistic That He's Being Compared to PT Barnum', *Business Insider*, 23 April 2019 <https://www.businessinsider.com/tesla-robo-taxis-elon-musk-pt-barnum-circus-2019-4> [accessed 11 January 2021].

24 Andrew Barclay, 'Why Is It So Hard to Make a Truly Self-Driving Car?', *South China Morning Post*, 5 July 2018 <https://www.scmp.com/abacus/tech/article/3028605/why-it-so-hard-make-truly-self-driving-car> [accessed 11 January 2021].

25 Rani Molla, 'How Apple's iPhone Changed the World: 10 Years in 10 Charts', *Vox*, 26 June 2017 <https://www.vox.com/2017/6/26/15821652/iphone-apple-10-year-anniversary-launch-mobile-stats-smart-phone-steve-jobs> [accessed 22 July 2020].

26 Ritwik Banerjee, Joydeep Bhattacharya and Priyama Majumdar, 'Exponential-Growth Prediction Bias and Compliance with Safety Measures Related to COVID-19', *Social Science & Medicine*, 268, January 2021, 113473 <https://doi.org/10.1016/j.socscimed.2020.113473>.

27 Robert C. Allen, 'Engels' Pause: Technical Change, Capital Accumulation, and Inequality in the British Industrial Revolution', *Explorations in Economic History*, 46(4) 2009, pp. 418–435 <https://doi.org/10.1016/j.eeh.2009.04.004>.

28 Hans-Joachim Voth, 'The Longest Years: New Estimates of Labor Input in England, 1760–1830', *The Journal of Economic History*, 61(4), December 2001, pp. 1065–1082 <https://doi.org/10.1017/S0022050701042085>. See also Charlie Giattino, Esteban Ortiz-Ospina and Max Roser, 'Working Hours', *Our World in Data*, 2013 <https://ourworldindata.org/working-hours> [accessed 25 February 2021].

29 Don Manuel Alvarez Espriella (pseudonym of Robert Southey), *Letters from England*, Letter XXXVI, 'Thursday July 7 – Birmingham 1814', p. 56.

30 Friedrich Engels, *The Condition of the Working-Class in England* (Panther, 1969) <https://www.marxists.org/archive/marx/works/1845/condition-working-class/> [accessed 18 December 2020].

31 Tony Blair, conversation with the author, London, 2019.

32 Anthony Giddens, *The Constitution of Society: Outline of the Theory of Structuration* (Berkeley, CA: University of California Press, 1986), p. 24.

33 Richard R. Nelson, *Technology, Institutions, and Economic Growth* (Cambridge, MA: Harvard University Press, 2005), p. 153.

34 Reuters, 'Galileo "Heresies" Still Under Study, Pope Says', *New York Times*, 10 May 1983 <https://www.nytimes.com/1983/05/10/world/galileo-heresies-still-under-study-pope-says.html> [accessed 24 July 2020].

35 'Vatican Admits Galileo Was Right', *New Scientist*, 7 November 1992 <https://www.newscientist.com/article/mg13618460-600-vatican-admits-galileo-was-right/> [accessed 24 July 2020].

36 Gareth A. Lloyd and Steven J. Sasson, 'Electronic Still Camera', US Patent No. 4131919A, 1977 <https://patents.google.com/patent/US4131919/en> [accessed 31 July 2020].

37 James Estrin, 'Kodak's First Digital Moment', *Lens* [blog], *New York Times*, 12 August 2015 <https://lens.blogs.nytimes.com/2015/08/12/kodaks-first-digital-moment/> [accessed 31 July 2020].

38 Scott D. Anthony, 'Kodak's Downfall Wasn't About Technology', *Harvard Business Review*, 15 July 2016 <https://hbr.org/2016/07/kodaks-downfall-wasnt-about-technology> [accessed 14 December 2020].

39 Wire Staff, 'May 26, 1995: Gates, Microsoft Jump on "Internet Tidal Wave"', *Wired*, 2 May 2010 <https://www.wired.com/2010/05/0526bill-gates-internet-memo/> [accessed 25 February 2021].

40 Joel Hruska, 'Ballmer: IPhone Has "No Chance" of Gaining Significant Market Share', *Ars Technica*, 30 April 2007 <https://arstechnica.com/information-technology/2007/04/ballmer-says-iphone-has-no-chance-to-gain-significant-market-share/> [accessed 7 January 2021].

41 Chris Smith, 'Steve Ballmer Finally Explains Why He Thought the IPhone Would Be a Flop', *BGR*, 4 November 2016 <https://bgr.com/2016/11/04/ballmer-iphone-quote-explained/> [accessed 7 January 2021].

42 W. F. Ogburn, *Social Change with Respect to Culture and Original Nature* (New York: B. W. Huebsch, 1923), pp. 200–236.

43 Amanda Lenhart, 'Chapter 3: Video Games Are Key Elements in Friendships for Many Boys', in *Teens, Technology & Friendships* (Pew Research

Center: Internet & Technology, 6 August 2015) <https://www.pewresearch.org/internet/2015/08/06/chapter-3-video-games-are-key-elements-in-friendships-for-many-boys/> [accessed 25 February 2021].

44 Douglass C. North, *Institutions, Institutional Change and Economic Performance* (Cambridge University Press, 1990) <https://doi.org/10.1017/CBO9780511808678>.

45 '"No Place for Discontent": A History of the Family Dinner in America', *NPR*, 16 February 2016 <https://www.npr.org/sections/thesalt/2016/02/16/459693979/no-place-for-discontent-a-history-of-the-family-dinner-in-america> [accessed 26 March 2021].

46 Kathleen Thelen, *How Institutions Evolve: The Political Economy of Skills in Germany, Britain, the United States, and Japan* (Cambridge, UK: Cambridge University Press, 2004) <https://doi.org/10.1017/CBO9780511790997>.

CHAPTER 4: THE UNLIMITED COMPANY

1 R. H. Coase, 'The Nature of the Firm', *Economica*, 4(16), November 1937, pp. 386–405 <https://doi.org/10.1111/j.1468-0335.1937.tb00002.x>.

2 Geoffrey B. West, *Scale: The Universal Laws of Growth, Innovation, Sustainability, and the Pace of Life in Organisms, Cities, Economies, and Companies* (New York: Penguin Press, 2017), pp. 387–410.

3 Myoung Cha and Flora Yu, 'Pharma's First-to-Market Advantage', McKinsey & Company, 1 September 2014 <https://www.mckinsey.com/industries/pharmaceuticals-and-medical-products/our-insights/pharmas-first-to-market-advantage> [accessed 23 September 2020].

4 'Car Registrations', SMMT <https://www.smmt.co.uk/vehicle-data/car-registrations/> [accessed 23 September 2020].

5 'TV Manufacturers: LCD TV Market Share Worldwide 2018', Statista <https://www.statista.com/statistics/267095/global-market-share-of-lcd-tv-manufacturers/> [accessed 23 September 2020].

6 'Standard Ogre', *The Economist*, 23 December 1999 <https://www.economist.com/business/1999/12/23/standard-ogre> [accessed 23 September 2020].

7 'Teen IPhone Ownership Hits an All-Time High in Long-Running Survey', MacRumors, 8 April 2020 <https://www.macrumors.com/2020/04/08/teen-iphone-ownership-hits-all-time-high/> [accessed 21 April 2021].

8 Keach Hagey and Suzanne Vranica, 'How Covid-19 Supercharged the Advertising "Triopoly" of Google, Facebook and Amazon', *Wall Street Journal*,

19 March 2021 <https://www.wsj.com/articles/how-covid-19-supercharged-the-advertising-triopoly-of-google-facebook-and-amazon-11616163738> [accessed 13 April 2021].

9 Liyin Yeo, 'The U.S. Rideshare Industry: Uber vs. Lyft', *Bloomberg Second Measure*, 2020 <https://secondmeasure.com/datapoints/rideshare-industry-overview/> [accessed 23 September 2020].

10 John Van Reenen and Christina Patterson, 'Research: The Rise of Superstar Firms Has Been Better for Investors than for Employees', *Harvard Business Review*, 11 May 2017 <https://hbr.org/2017/05/research-the-rise-of-superstar-firms-has-been-better-for-investors-than-for-employees> [accessed 2 September 2020].

11 James Manyika et al., *"Superstars": The Dynamics of Firms, Sectors, and Cities Leading the Global Economy* (McKinsey & Company, 24 October 2018) <https://www.mckinsey.com/featured-insights/innovation-and-growth/superstars-the-dynamics-of-firms-sectors-and-cities-leading-the-global-economy> [accessed 19 December 2020].

12 David Autor et al., 'The Fall of the Labor Share and the Rise of Superstar Firms', *The Quarterly Journal of Economics*, 135(2), May 2020, pp. 645–709 <https://doi.org/10.1093/qje/qjaa004>.

13 'Companies – The Rise of the Superstars', *The Economist*, 15 September 2016 <https://www.economist.com/special-report/2016/09/15/the-rise-of-the-superstars> [accessed 2 September 2020].

14 'What Drives Productivity Growth?', United States Census Bureau <https://www.census.gov/library/stories/2019/10/what-drives-productivity-growth.html> [accessed 2 September 2020].

15 *Annual Report of the Directors of American Telephone & Telegraph Company to the Stockholders* (Boston: Geo H. Ellis Co., 1908), p. 23.

16 The canonical description of this is in Michael E. Porter, *Competitive Advantage: Creating and Sustaining Superior Performance* (New York: Free Press, 1998).

17 Laure Claire Reillier and Benoit Reillier, *Platform Strategy: How to Unlock the Power of Communities and Networks to Grow Your Business* (London: Routledge, 2017), p. 4.

18 'EBay's US Sales Grow 22 per cent in 2020 Adding 11 Million New Customers', *Digital Commerce 360*, 17 February 2021 <https://www.digitalcommerce360.com/article/ebays-sales/> [accessed 28 March 2021].

19 'Alibaba: Cumulative Active Online Buyers 2020', Statista <https://www.statista.com/statistics/226927/alibaba-cumulative-active-online-buyers-taobao-tmall/> [accessed 28 March 2021].

20 Alex Sherman, 'TikTok Reveals Detailed User Numbers for the First Time', *CNBC*, 24 August 2020 <https://www.cnbc.com/2020/08/24/tiktok-reveals-us-global-user-growth-numbers-for-first-time.html> [accessed 28 March 2021].

21 Sangeet Paul Choudary, *Platform Scale: How an Emerging Business Model Helps Startups Build Large Empires with Minimum Investment* (Singapore: Platform Thinking Labs Pte. Ltd, 2015) p. 36.

22 The best book on the intangible economy is Jonathan Haskel and Stian Westlake, *Capitalism without Capital: The Rise of the Intangible Economy* (Princeton, NJ: Princeton University Press, 2018).

23 *Intangible Asset Market Value Study*, Ocean Tomo <https://www.oceantomo.com/intangible-asset-market-value-study/> [accessed 27 August 2020].

24 Stian Westlake, quoted in Azeem Azhar, 'Understanding the Intangible Economy', *Exponential View*, 5 July 2019 <https://www.exponentialview.co/p/capitalism-without-capital> [accessed 28 August 2020].

25 *World Intellectual Property Report 2017: Intangible Capital in Global Value Chains* (World Intellectual Property Organization, 2017).

26 Barney Pell, personal conversation with the author, June 2015.

27 Matt Turck, 'The Power of Data Network Effects', *Matt Turck* [blog], 2016 <https://mattturck.com/the-power-of-data-network-effects/> [accessed 3 August 2020].

28 Seyed M. Mirtaheri et al., 'A Brief History of Web Crawlers', in *CASCON '13: Proceedings of the 2013 Conference of the Center for Advanced Studies on Collaborative Research* (Toronto: IBM Corp., 2013), pp. 40–54.

29 Tim O'Reilly, 'Network Effects in Data', *O'Reilly Radar*, 27 October 2008 <http://radar.oreilly.com/2008/10/network-effects-in-data.html> [accessed 9 December 2020].

30 West, *Scale*, p. 393.

31 Author's analysis of various company disclosures.

32 'Amount of Original Content Titles on Netflix 2019', Statista <https://www.statista.com/statistics/883491/netflix-original-content-titles/> [accessed 30 March 2021]; Gavin Bridge, 'Netflix Released More Originals in 2019 Than the Entire TV Industry Did in 2005', *Variety*, 17 December 2019 <https://

variety.com/2019/tv/news/netflix-more-2019-originals-than-entire-tv-industry-in-2005-1203441709/> [accessed 30 March 2021].

33 W. Brian Arthur, 'Increasing Returns and the New World of Business', *Harvard Business Review*, 1 July 1996 <https://hbr.org/1996/07/increasing-returns-and-the-new-world-of-business> [accessed 31 July 2020].

34 Peter Thiel, 'Competition Is for Losers', *Wall Street Journal*, 12 September 2014 <https://online.wsj.com/articles/peter-thiel-competition-is-for-losers-1410535536> [accessed 9 October 2020].

35 CogX, 'Bringing Inclusive Financial Services to the World', 15 June 2018 <https://www.youtube.com/watch?v=m0YT4O4CWG4> [accessed 29 December 2020].

36 C. K. Prahalad and Gary Hamel, 'The Core Competence of the Corporation', *Harvard Business Review*, 1 May 1990 <https://hbr.org/1990/05/the-core-competence-of-the-corporation> [accessed 27 August 2020].

37 Elizabeth Gibney, 'Google Revives Controversial Cold-Fusion Experiments', *Nature*, 569(7758), 27 May 2019, p. 611 <https://doi.org/10.1038/d41586-019-01683-9>.

38 'Apple Announces App Store Small Business Program', Apple Newsroom, November 2020 <https://www.apple.com/newsroom/2020/11/apple-announces-app-store-small-business-program/> [accessed 29 December 2020].

39 Austen Goslin, 'Why Fortnite Is the Most Important Game of the Decade', *Polygon*, 14 November 2019 <https://www.polygon.com/2019/11/14/20965516/fortnite-battle-royale-most-important-game-2010s> [accessed 29 December 2020].

40 'Antitrust: Google Fined €1.49 Billion for Online Advertising Abuse', European Commission, 20 March 2019 <https://ec.europa.eu/commission/presscorner/detail/en/IP_19_1770> [accessed 30 March 2021].

41 Dimitrios Katsifis, 'The CMA Publishes Final Report on Online Platforms and Digital Advertising', *The Platform Law Blog*, 6 July 2020 <https://theplatformlaw.blog/2020/07/06/the-cma-publishes-final-report-on-online-platforms-and-digital-advertising/> [accessed 30 March 2021].

42 Yoram Wijngaard, personal correspondence between Dealroom and the author, 30 March 2021.

43 Dashun Wang and James A. Evans, 'Research: When Small Teams Are Better than Small Ones', *Harvard Business Review*, 21 January 2019 <https://hbr.org/2019/02/research-when-small-teams-are-better-than-big-ones>.

44 Joel Klinger, Juan Mateos-Garcia and Konstantinos Stathoulopoulos, 'A Narrowing of AI Research?', arXiv:2009.10385 [cs.CY], 2020 <http://arxiv.org/abs/2009.10385> [accessed 30 March 2021].

45 Michael Gofman and Zhao Jin, 'Artificial Intelligence, Human Capital, and Innovation', *SSRN Electronic Journal*, October 2019 <https://doi.org/10.2139/ssrn.3449440>.

46 Azeem Azhar, 'Beneficial Artificial Intelligence: My Conversation with Stuart Russell', 22 August 2019 <https://www.exponentialview.co/p/-beneficial-artificial-intelligence> [accessed 16 April 2021].

47 Gofman and Jin, 'Artificial Intelligence'.

48 Klinger, Mateos-Garcia and Stathoulopoulos, 'A Narrowing of AI Research?'.

49 'A Tech CEO's Guide to Taxes', *The Economist*, 9 January 2021 <https://www.economist.com/business/2021/01/09/a-tech-ceos-guide-to-taxes> [accessed 21 March 2021].

50 Vanessa Houlder, 'Q&A: What Is the Double Irish?', 2014 <https://www.ft.com/content/f7a2b958-4fc8-11e4-908e-00144feab7de> [accessed 24 March 2021].

51 Richard Waters, 'Google to End Use of "Double Irish" as Tax Loophole Set to Close', 2020 <https://www.ft.com/content/991f11ae-2c51-11ea-bc77-65e4aa615551> [accessed 24 March 2021].

52 Sanjana Varghese, 'This Is Why Silicon Valley Giants Are Paying Such Low Taxes in the UK', *Wired*, 12 October 2018 <https://www.wired.co.uk/article/facebook-uk-tax-bill> [accessed 14 October 2020].

53 Lina M. Khan, 'Amazon's Antitrust Paradox', *Yale Law Journal*, 126(3), January 2017 <https://www.yalelawjournal.org/note/amazons-antitrust-paradox> [accessed 22 April 2020].

54 Author's analysis of various issues of the Ofcom *Communications Market Report*.

55 'The Digital Markets Act: Ensuring Fair and Open Digital Markets', European Commission <https://ec.europa.eu/info/strategy/priorities-2019-2024/europe-fit-digital-age/digital-markets-act-ensuring-fair-and-open-digital-markets_en> [accessed 10 January 2021].

56 'BT Agrees to Legal Separation of Openreach', Ofcom, 10 March 2017 <https://www.ofcom.org.uk/about-ofcom/latest/media/media-releases/2017/bt-agrees-to-legal-separation-of-openreach> [accessed 10 January 2021].

57 'The Digital Services Act: Ensuring a Safe and Accountable Online Environment', European Commission <https://ec.europa.eu/info/strategy/priorities-2019-2024/europe-fit-digital-age/digital-services-act-ensuring-safe-and-accountable-online-environment_en> [accessed 29 March 2021].

CHAPTER 5: LABOUR'S LOVES LOST

1 Larry Elliott, 'Millions of UK Workers at Risk of Being Replaced by Robots', *The Guardian*, 24 March 2017 <https://www.theguardian.com/technology/2017/mar/24/millions-uk-workers-risk-replaced-robots-study-warns>.

2 Alex Williams, 'Will Robots Take Our Children's Jobs?', *New York Times*, 11 December 2017 <https://www.nytimes.com/2017/12/11/style/robots-jobs-children.html>.

3 'Labor: Sabotage at Lordstown?', *Time*, 7 February 1972 <http://content.time.com/time/subscriber/article/0,33009,905747,00.html> [accessed 3 April 2021].

4 John Maynard Keynes, 'Economic Possibilities for Our Grandchildren', in *Essays in Persuasion* (London: Palgrave Macmillan UK, 2010), pp. 321–332 <https://doi.org/10.1007/978-1-349-59072-8_25>.

5 Creative Destruction Lab, 'Geoff Hinton: On Radiology', 24 November 2016 <https://www.youtube.com/watch?v=2HMPRXstSvQ> [accessed 24 February 2021].

6 Paul Daugherty, H. James Wilson and Paul Michelman, 'Revisiting the Jobs That Artificial Intelligence Will Create', *MIT Sloan Management Review* (Summer 2017).

7 Lana Bandoim, 'Robots Are Cleaning Grocery Store Floors During the Coronavirus Outbreak', *Forbes*, 8 April 2020 <https://www.forbes.com/sites/lanabandoim/2020/04/08/robots-are-cleaning-grocery-store-floors-during-the-coronavirus-outbreak/> [accessed 24 February 2021].

8 Jame DiBiasio, 'A.I. Drives China Techfins into Car Insurance – and Beyond', *Digital Finance*, 3 June 2020 <https://www.digfingroup.com/insurtech-ai/> [accessed 24 February 2021].

9 Carl Benedikt Frey and Michael Osborne, 'The Future of Employment: How Susceptible Are Jobs to Computerisation?', Oxford Martin School working paper, 17 September 2013 <https://www.oxfordmartin.ox.ac.uk/publications/the-future-of-employment.pdf> [accessed 14 September 2020].

10 Google Scholar citation search shows 7,554 citations to the Frey and Osborne research as of 21 February 2021 <https://scholar.google.com/scholar?cites=8817314921922525274&as_sdt=2005&sciodt=0,5&hl=en>.

11 'Forrester Predicts Automation Will Displace 24.7 Million Jobs and Add 14.9 Million Jobs By 2027', Forrester, 3 April 2017 <https://go.forrester.com/press-newsroom/forrester-predicts-automation-will-displace-24-7-million-jobs-and-add-14-9-million-jobs-by-2027/> [accessed 14 September 2020].

12 'Robots "to Replace 20 Million Factory Jobs" ', *BBC News*, 26 June 2019 <https://www.bbc.com/news/business-48760799> [accessed 20 September 2020].

13 John Detrixhe, 'Deutsche Bank's CEO Hints Half Its Workers Could Be Replaced by Machines', *Quartz*, 8 July 2019 <https://qz.com/1123703/deutsche-bank-ceo-john-cryan-suggests-half-its-workers-could-be-replaced-by-machines/> [accessed 25 September 2020].

14 Eric J. Savitz, 'UiPath Stock Has Strong Debut, Stock Soars 25 per cent From IPO Price', *Barron's*, 21 April 2021 <https://www.barrons.com/articles/uipath-stock-has-strong-debut-stock-soars-25-from-ipo-price-51619031782> [accessed 29 April 2021].

15 U.S. Bureau of Labor Statistics, 'All Employees, Manufacturing', *FRED, Federal Reserve Bank of St. Louis* (FRED, Federal Reserve Bank of St. Louis, 1939) <https://fred.stlouisfed.org/series/MANEMP> [accessed 26 August 2020]. This represented about one in five of all non-agricultural American workers.

16 'Productivity per Hour Worked', *Our World in Data* <https://ourworldindata.org/grapher/labor-productivity-per-hour-PennWorldTable> [accessed 26 August 2020].

17 Mark Muro, 'It Won't Be Easy to Bring Back Millions of Manufacturing Jobs', Brookings, 18 November 2016 <https://www.brookings.edu/blog/the-avenue/2016/11/18/it-wont-be-easy-to-bring-back-millions-of-manufacturing-jobs/> [accessed 27 August 2020].

18 Agis Salpukas, 'General Motors Reports '80 Loss of $763 Million', *New York Times*, 3 February 1981 <https://www.nytimes.com/1981/02/03/business/general-motors-reports-80-loss-of-763-million.html> [accessed 26 August 2020].

19 One popular practice in Exponential Age firms is to use large numbers of contractors and temporary workers alongside full-time employees. In Google's case, this 'shadow work force' was about the same size as the

full-time employee base. However, this doesn't fundamentally challenge my argument. (See Daisuke Wakabayashi, 'Google's Shadow Work Force: Temps Who Outnumber Full-Time Employees', *The New York Times*, 28 May 2019, section Technology <https://www.nytimes.com/2019/05/28/technology/google-temp-workers.html> [accessed 17 May 2021].)

20 'Employment – Employment Rate – OECD Data', *TheOECD* <http://data.oecd.org/emp/employment-rate.htm> [accessed 14 September 2020].

21 'Unemployment, Total (per cent of Total Labor Force)', World Bank Data <https://data.worldbank.org/indicator/SL.UEM.TOTL.ZS> [accessed 14 September 2020].

22 Hans P. Moravec, *Mind Children: The Future of Robot and Human Intelligence* (Cambridge, MA: Harvard University Press, 1988), p. 15.

23 'Liquidity, Volatility, Fragility', *Goldman Sachs Global Macro Research*, 68, June 2018.

24 John Gittelson, 'End of Era: Passive Equity Funds Surpass Active in Epic Shift', *Bloomberg*, 11 September 2019 <https://www.bloomberg.com/news/articles/2019-09-11/passive-u-s-equity-funds-eclipse-active-in-epic-industry-shift> [accessed 14 October 2020].

25 'March of the Machines – The Stockmarket Is Now Run by Computers, Algorithms and Passive Managers', *The Economist*, 5 October 2019 <https://www.economist.com/briefing/2019/10/05/the-stockmarket-is-now-run-by-computers-algorithms-and-passive-managers> [accessed 14 October 2020].

26 Michael Polanyi and Amartya Sen, *The Tacit Dimension* (Chicago, IL: University of Chicago Press, 2009), p. 4.

27 David Graeber, *Bullshit Jobs: A Theory* (New York: Simon & Schuster, 2018), p. 236

28 Carl Benedikt Frey, *The Technology Trap: Capital, Labor, and Power in the Age of Automation* (Princeton, NJ: Princeton University Press, 2019), p. 311.

29 Staci D. Kramer, 'The Biggest Thing Amazon Got Right: The Platform', *Gigaom*, 12 October 2011 <https://gigaom.com/2011/10/12/419-the-biggest-thing-amazon-got-right-the-platform/> [accessed 18 September 2020].

30 The approach has become de rigueur among other digital cognoscenti, but Bezos's email should surely be considered as one of the single most important internal communications of all time.

31 Chris Johnston, 'Amazon Opens a Supermarket with No Checkouts', *BBC News*, 22 January 2018 <https://www.bbc.com/news/business-42769096> [accessed 18 September 2020].

32 Peter Holley, 'Amazon's One-Day Delivery Service Depends on the Work of Thousands of Robots', *Washington Post*, 7 June 2019 <https://www.washingtonpost.com/technology/2019/06/07/amazons-one-day-delivery-service-depends-work-thousands-robots/> [accessed 18 September 2020].

33 Harry Dempsey, 'Amazon to Hire Further 100,000 Workers in US and Canada', 14 September 2020 <https://www.ft.com/content/9817aae3-1e89-4383-aa34-742447d5794a> [accessed 18 September 2020].

34 'Netflix Continues to Hire Through the Pandemic, Says Co-CEO Reed Hastings', *Bloomberg*, 9 September 2020 <https://www.bloomberg.com/news/videos/2020-09-09/netflix-continues-to-hire-through-the-pandemic-video> [accessed 18 September 2020].

35 'Netflix: Number of Employees 2006-2020', Macro Trends <https://www.macrotrends.net/stocks/charts/NFLX/netflix/number-of-employees> [accessed 27 March 2021].

36 Vishnu Rajamanickm, 'JD.Com Opens Automated Warehouse That Employs Four People but Fulfills 200,000 Packages Daily', *FreightWaves*, 25 June 2018 <https://www.freightwaves.com/news/technology/jdcom-opens-automated-warehouse-that-employs-four-people-but-fulfills-200000-packages-daily> [accessed 27 March 2021].

37 Reuters Staff, 'Dish to Close 300 Blockbuster Stores, 3,000 Jobs May Be Lost', *Reuters*, 23 January 2013 <https://www.reuters.com/article/us-blockbuster-storeclosings-IDUSBRE90M05I20130123> [accessed 7 January 2021].

38 Daron Acemoglu, Claire LeLarge and Pascual Restrepo, *Competing with Robots: Firm-Level Evidence from France*, Working Paper Series (National Bureau of Economic Research, February 2020) <https://doi.org/10.3386/w26738>.

39 Daron Acemoglu and Pascual Restrepo, *Robots and Jobs: Evidence from US Labor Markets*, Working Paper Series (National Bureau of Economic Research, March 2017) <https://doi.org/10.3386/w23285>.

40 David Klenert, Enrique Fernández-Macías and José-Ignacio Antón, 'Don't Blame It on the Machines: Robots and Employment in Europe', *VoxEU*, 24 February 2020 <https://voxeu.org/article/dont-blame-it-machines-robots-and-employment-europe> [accessed 10 September 2020].

41 Till Leopold et al., *The Future of Jobs 2018*, World Economic Forum <https://wef.ch/2NH6NiV> [accessed 25 September 2020].

42 Leslie Willcocks, 'Robo-Apocalypse Cancelled? Reframing the Automation and Future of Work Debate', *Journal of Information Technology*, 35(4), 2020, pp. 286–302 <https://doi.org/10.1177/0268396220925830>.

43 Chris Urmson, personal correspondence with the author, 26 February 2021.

44 'Will AI Destroy More Jobs Than It Creates Over the Next Decade?', *Wall Street Journal*, 1 April 2019 <https://www.wsj.com/articles/will-ai-destroy-more-jobs-than-it-creates-over-the-next-decade-11554156299> [accessed 11 January 2021].

45 'Company Information', Uber Newsroom Pakistan <https://www.uber.com/en-PK/newsroom/company-info/> [accessed 21 September 2020].

46 'Mechanical Turk: Research in the Crowdsourcing Age', *Pew Research Center: Internet & Technology*, 11 July 2016 <https://www.pewresearch.org/internet/2016/07/11/research-in-the-crowdsourcing-age-a-case-study/> [accessed 28 September 2020].

47 Jeff Howe, 'The Rise of Crowdsourcing', *Wired*, 1 June 2006 <https://www.wired.com/2006/06/crowds/> [accessed 28 September 2020].

48 Nicole Lyn Pesce, 'This Chart Shows How Uber Rides Sped Past NYC Yellow Cabs in Just Six Years', *MarketWatch*, 9 August 2019 <https://www.marketwatch.com/story/this-chart-shows-how-uber-rides-sped-past-nyc-yellow-cabs-in-just-six-years-2019-08-09> [accessed 7 January 2021].

49 Kelle Howson et al., 'Platform Workers, the Future of Work and Britain's Election', *Media@LSE*, 11 December 2019 <https://blogs.lse.ac.uk/medialse/2019/12/11/platform-workers-the-future-of-work-and-britains-election/> [accessed 7 January 2021].

50 James Manyika et al., *Connecting Talent with Opportunity in the Digital Age* (McKinsey & Company, 1 June 2015) <https://www.mckinsey.com/featured-insights/employment-and-growth/connecting-talent-with-opportunity-in-the-digital-age> [accessed 6 October 2020].

51 Neil Munshi, 'Tech Start-Ups Drive Change for Nigerian Truckers', *Financial Times*, 26 August 2019 <https://www.ft.com/content/c6a3d1f2-c27d-11e9-a8e9-296ca66511c9> [accessed 21 September 2020].

52 'Upwork Reports Fourth Quarter and Full Year 2020 Financial Results', Upwork Inc., 23 February 2021 <https://investors.upwork.com/news-releases/news-release-details/upwork-reports-fourth-quarter-and-full-year-2020-financial> [accessed 21 April 2021].

53 Lijin Yeo, 'The U.S. Rideshare Industry: Uber vs. Lyft', *Bloomberg Second Measure*,2020<https://secondmeasure.com/datapoints/rideshare-industry-overview/> [accessed 23 September 2020].

54 'Gig Economy Research', Gov.uk, 7 February 2018 <https://www.gov.uk/government/publications/gig-economy-research> [accessed 21 September 2020].

55 Ravi Agrawal, 'The Hidden Benefits of Uber', *Foreign Policy*, 16 July 2018 <https://foreignpolicy.com/2018/07/16/why-india-gives-uber-5-stars-gig-economy-jobs/> [accessed 21 September 2020].

56 Department for Business, Energy & Industrial Strategy, 'Gig Economy Research', Gov.uk, 7 February 2018 <https://www.gov.uk/government/publications/gig-economy-research> [accessed 21 September 2020].

57 Directorate General for Internal Policies, *The Social Protection of Workers in the Platform Economy, Study for the EMPL Committee, IP/A/EMPL/2016-11* (European Parliament, 2017).

58 Nicole Karlis, 'DoorDash Drivers Make an Average of $1.45 an Hour, Analysis Finds', *Salon*, 19 January 2020 <https://www.salon.com/2020/01/19/doordash-drivers-make-an-average-of-145-an-hour-analysis-finds/> [accessed 27 March 2021].

59 Kate Conger, 'Uber and Lyft Drivers in California Will Remain Contractors', *New York Times*, 4 November 2020 <https://www.nytimes.com/2020/11/04/technology/california-uber-lyft-prop-22.html> [accessed 12 January 2021].

60 Mary-Ann Russon, 'Uber Drivers Are Workers Not Self-Employed, Supreme Court Rules', *BBC News*, 19 February 2021 <https://www.bbc.com/news/business-56123668> [accessed 29 March 2021].

61 'Judgement: Uber BV and Others (Appellants) v Aslam and Others (Respondents)', 19 February 2021 <https://www.supremecourt.uk/cases/docs/uksc-2019-0029-judgment.pdf> [accessed 19 March 2021].

62 'Frederick Winslow Taylor: Father of Scientific Management Thinker', The British Library <https://www.bl.uk/people/frederick-winslow-taylor> [accessed 29 March 2021].

63 Nikil Saval, *Cubed: A Secret History of the Workplace* (New York: Anchor Books, 2015), p. 42.

64 Saval, *Cubed*, p. 56.

65 Alex Rosenblat, Tamara Kneese and danah boyd, *Workplace Surveillance* (Data & Society Research Institute, 4 January 2017) <https://doi.org/10.31219/osf.io/7ryk4>.

66 'In March 2017, the Japanese Government Formulated the Work Style Reform Action Plan.', *Social Innovation*, September 2017 <https://social-innovation.hitachi/en/case_studies/ai_happiness/> [accessed 6 October 2020].

67 Alex Hern, 'Microsoft Productivity Score Feature Criticised as Workplace Surveillance', *The Guardian*, 26 November 2020 <http://www.theguardian.

com/technology/2020/nov/26/microsoft-productivity-score-feature-criticised-workplace-surveillance> [accessed 1 April 2021].

68 Stephen Chen, 'Chinese Surveillance Programme Mines Data from Workers' Brains', *South China Morning Post*, 28 April 2018 <https://www.scmp.com/news/china/society/article/2143899/forget-facebook-leak-china-mining-data-directly-workers-brains> [accessed 6 October 2020].

69 Robert Booth, 'Unilever Saves on Recruiters by Using AI to Assess Job Interviews', *The Guardian*, 25 October 2019 <http://www.theguardian.com/technology/2019/oct/25/unilever-saves-on-recruiters-by-using-ai-to-assess-job-interviews> [accessed 6 October 2020].

70 Chartered Institute of Personnel and Development, 'Workplace Technology: The Employee Experience' (CIPD: July 2020) <https://www.cipd.co.uk/Images/workplace-technology-1_tcm18-80853.pdf> [accessed 19 May 2021].

71 Sarah O'Connor, 'When Your Boss Is an Algorithm', *Financial Times*, 7 September 2016 <https://www.ft.com/content/88fdc58e-754f-11e6-b60a-de4532d5ea35> [accessed 3 August 2020].

72 Tom Barratt et al., 'Algorithms Workers Can't See Are Increasingly Pulling the Management Strings', *Management Today*, 25 August 2020 <http://www.managementtoday.co.uk/article/1692636?utm_source=website&utm_medium=social> [accessed 3 April 2021].

73 Michael Sainato, ' "I'm Not a Robot": Amazon Workers Condemn Unsafe, Grueling Conditions at Warehouse', *The Guardian*, 5 February 2020 <http://www.theguardian.com/technology/2020/feb/05/amazon-workers-protest-unsafe-grueling-conditions-warehouse> [accessed 6 October 2020].

74 Colin Lecher, 'How Amazon Automatically Tracks and Fires Warehouse Workers for "Productivity" ', *The Verge*, 25 April 2019 <https://www.theverge.com/2019/4/25/18516004/amazon-warehouse-fulfillment-centers-productivity-firing-terminations> [accessed 6 October 2020].

75 James Manyika et al., *A New Look at the Declining Labor Share of Income in the United States* (McKinsey Global Institute, May 2019), p. 64.

76 'The Productivity–Pay Gap', Economic Policy Institute, July 2019 <https://www.epi.org/productivity-pay-gap/> [accessed 22 April 2021].

77 Manayika et al., *A New Look at the Declining Labor Share of Income in the United States*, p. 64.

78 Autor et al., *The Fall of the Labor Share and the Rise of Superstar Firms*, p. 106.

79 Rani Molla, 'Facebook, Google and Netflix Pay a Higher Median Salary than Exxon, Goldman Sachs or Verizon', *Vox*, 30 April 2018 <https://www.vox.com/2018/4/30/17301264/how-much-twitter-google-amazon-highest-paying-salary-tech> [accessed 18 October 2020].

80 'Zymergen: Case Study', The Partnership on AI <https://www.partnershiponai.org/case-study/zymergen/> [accessed 18 October 2020].

81 Frey, *The Technology Trap*, pp. 137–139.

82 Robert C. Allen, 'Engels' Pause: Technical Change, Capital Accumulation, and Inequality in the British Industrial Revolution', *Explorations in Economic History*, 46(4), 2009, pp. 418–435 <https://doi.org/10.1016/j.eeh.2009.04.004>.

83 John Maynard Keynes, *A Tract on Monetary Reform* (London: Macmillan, 1923), p. 80.

84 Delphine Strauss and Siddharth Venkataramakrishnan, 'Dutch Court Rulings Break New Ground on Gig Worker Data Rights', *Financial Times*, 12 March 2021 <https://www.ft.com/content/334d1ca5-26af-40c7-a9c5-c76e3e57fba1> [accessed 16 April 2021].

85 Emma Peaslee, 'Results from the City That Just Gave Away Cash', *NPR*, 9 March 2021 <https://www.npr.org/sections/money/2021/03/09/975009239/results-from-the-city-that-just-gave-away-cash> [accessed 3 April 2021].

86 Neil Lee and Stephen Clarke, 'Do Low-Skilled Workers Gain from High-Tech Employment Growth? High-Technology Multipliers, Employment and Wages in Britain', *Research Policy*, 48(9), November 2019, 103803 <https://doi.org/10.1016/j.respol.2019.05.012>.

87 'Trade Union', OECD.Stat <https://stats.oecd.org/Index.aspx?DataSetCode=TUD>.

88 Ben Tarnoff, 'The Making of the Tech Worker Movement', *Logic Magazine*, 4 May 2020 <https://logicmag.io/the-making-of-the-tech-worker-movement/full-text/> [accessed 3 April 2021].

89 Irina Ivanova, 'Amazon Picks Twitter Fight with Bernie Sanders and Elizabeth Warren amid Union Campaign', *CBS News*, 26 March 2021 <https://www.cbsnews.com/news/amazon-bernie-sanders-elizabeth-warren-union-vote/> [accessed 29 March 2021].

90 Bethan Staton, 'The Upstart Unions Taking on the Gig Economy and Outsourcing', *Financial Times*, 18 January 2020 <https://www.ft.com/content/576c68ea-3784-11ea-a6d3-9a26f8c3cba4> [accessed 12 January 2021].

91 'Table 5. Union Affiliation of Employed Wage and Salary Workers by State', US Bureau of Labour Statistics <https://www.bls.gov/news.release/union2.t05.htm> [accessed 1 April 2021].

92 'Technology May Help to Revive Organised Labour', *The Economist*, 15 November 2018 <https://www.economist.com/briefing/2018/11/15/technology-may-help-to-revive-organised-labour>.

CHAPTER 6: THE WORLD IS SPIKY

1 Coco Liu and Shunsuke Tabeta, 'China Car Startup Dodges Trump Tariffs with AI and 3D Printing', *Nikkei Asia*, 27 September 2019 <https://asia.nikkei.com/Spotlight/Startups-in-Asia/China-car-startup-dodges-Trump-tariffs-with-AI-and-3D-printing> [accessed 4 September 2020].

2 'Is Globalization an Engine of Economic Development?', *Our World in Data* <https://ourworldindata.org/is-globalization-an-engine-of-economic-development> [accessed 9 October 2020].

3 The phrase 'the world is spiky' was originally used by Richard Florida in his essay 'The World Is Spiky' in *Atlantic Monthly*, October 2005. Florida used it to describe the relative inequalities across and within countries and, in particular, the disproportionate importance of large cities in economic development. I use the term in a slightly different context, referring to the overall breakdown of the thesis that technology necessarily enables single large global markets, mediated by a set of common rules.

4 Philip Garnett, Bob Doherty, and Tony Heron, 'Vulnerability of the United Kingdom's Food Supply Chains Exposed by COVID-19', *Nature Food*, 1(6), 2020, pp. 315–318 <https://doi.org/10.1038/s43016-020-0097-7>.

5 Alex Lee, 'How the UK's Just-in-Time Delivery Model Crumbled under Coronavirus', *Wired*, 30 March 2020 <https://www.wired.co.uk/article/stockpiling-supermarkets-coronavirus> [accessed 11 September 2020].

6 UBS, *The Food Revolution*, July 2019 <https://www.ubs.com/global/en/ubs-society/our-stories/2019/future-of-food/_jcr_content/mainpar/toplevelgrid_1749059381/col1/linklist/link.1695495471.file/bGluay9w-YXRoPS9jb250ZW50L2RhbS91YnMvZ2xvYmFsL3Vicy1zb2Np-ZXR5LzIwMTkvZm9vZC1yZXZvbHV0aW9uLWp1bHkucGRm/food-revolution-july.pdf>.

7 'Growing Higher – New Ways to Make Vertical Farming Stack up', *The Economist*, 31 August 2019 <https://www.economist.com/science-and-

technology/2019/08/31/new-ways-to-make-vertical-farming-stack-up> [accessed 4 August 2020].

8 'World's Biggest Rooftop Greenhouse Opens in Montreal', Phys.org, 26 August 2020 <https://phys.org/news/2020-08-world-biggest-rooftop-greenhouse-montreal.html> [accessed 5 September 2020].

9 Elizabeth Curmi et al., 'Feeding the Future', *Citi Global Perspectives and Solutions*, November 2018 <https://www.citivelocity.com/citigps/feeding-the-future/> [accessed 18 March 2021].

10 Joel Jean, Patrick Richard Brown and Institute of Physics (Great Britain), *Emerging Photovoltaic Technologies* (Bristol, UK: IOP Publishing, 2020), pp. 1–5 <https://iopscience.iop.org/book/978-0-7503-2152-5> [accessed 12 October 2020].

11 Brendan Coyne, 'Vehicle-to-Grid: Are We Nearly There Yet?', *the energyst*, 12 April 2019 <https://theenergyst.com/evs-v2g-vehicle-to-grid-battery-storage-smartgrid/> [accessed 6 September 2020].

12 Guste, 'Moixa Secures £5M Investment to Drive Global Growth', Moixa, 22 July 2020 <https://www.moixa.com/press-release/moixa-secures-5m-investment-to-drive-global-growth/> [accessed 6 September 2020].

13 'What Smart Manufacturers Know About Bundling Products and Services', *Knowledge@Wharton*, 30 September 2019 <https://knowledge.wharton.upenn.edu/article/what-smart-manufacturers-know-about-bundling-products-and-services/> [accessed 7 September 2020].

14 Kif Leswing, 'The $999 IPhone X Is Estimated to Include Less than $400 in Parts', *Business Insider*, November 2017 <https://www.businessinsider.com/iphone-x-teardown-parts-cost-ihs-markit-2017-11> [accessed 7 September 2020].

15 Kathy Chu and Ellen Emmerentze Jervell, 'At Western Firms Like Adidas, Rise of the Machines Is Fueled by Higher Asia Wages', *Wall Street Journal*, 9 June 2016 <https://www.wsj.com/articles/rise-of-the-machines-fueled-by-higher-asia-manufacturing-wages-1465457460> [accessed 14 October 2020].

16 Marc Bain, 'Adidas Is Shutting down the Robotic Factories That Were Supposed to Be Its Future', *Quartz*, 11 November 2019 <https://qz.com/1746152/adidas-is-shutting-down-its-speedfactories-in-germany-and-the-us/> [accessed 14 October 2020].

17 Eric Dustman, Kareem Elwakil and Miguel Smart, *Metals 3D Printing: Closing the Cost Gap and Getting to Value* (PricewaterhouseCoopers,

2019) <https://www.strategyand.pwc.com/gx/en/insights/2019/metals-3D-printing.html> [accessed 7 September 2020].

18 Zoe Kleinman, 'Coronavirus: Can We 3D-Print Our Way out of the PPE Shortage?', *BBC News*, 9 April 2020 <https://www.bbc.co.uk/news/health-52201696> [accessed 14 October 2020].

19 Pric Dustman et al., *Metals 3D Printing*.

20 Javier Blas and Jack Farchy, *The World for Sale* (London: Penguin Books, 2021), p. 220.

21 Blas and Farchy, *The World for Sale*, p. 221.

22 Atif Kubursi, 'Oil Crash Explained: How Are Negative Oil Prices Even Possible?', *The Conversation*, 20 April 2020 <https://theconversation.com/oil-crash-explained-how-are-negative-oil-prices-even-possible-136829>.

23 '3D Printing: A Threat to Global Trade' <www.ingwb.com/insights/research/3d-printing-a-threat-to-global-trade> [accessed 28 December 2020].

24 Nick Carey, 'UPS, SAP team up for on-demand 3D printing network', *Reuters*, 18 May 2016 <https://de.reuters.com/article/us-united-parcel-sap-se-3dprinting-IDUSKCN0Y90BR> [accessed 7 September 2020].

25 T. X. Hannes, 'Will Technological Convergence Reverse Globalization?', National Defense University Press, 12 July 2016 <http://ndupress.ndu.edu/Media/News/News-Article-View/Article/834357/will-technological-convergence-reverse-globalization/> [accessed 6 September 2020].

26 Nick Butler, 'US Energy Independence Has Its Costs', *Financial Times*, 2019 <https://www.ft.com/content/20870c24-0b86-11ea-b2d6-9bf4d1957a67> [accessed 10 May 2021].

27 'Urban Population (per cent of Total Population)', World Bank Data <https://data.worldbank.org/indicator/SP.URB.TOTL.IN.ZS> [accessed 11 January 2021].

28 Azeem Azhar, 'Don't Call Time on the Megacity', *Exponential View*, 20 May 2020 <https://www.exponentialview.co/p/-dont-call-time-on-the-megacity> [accessed 13 January 2021].

29 Genevieve Giuliano, Sanggyun Kang and Quan Yuan, 'Agglomeration Economies and Evolving Urban Form', *The Annals of Regional Science*, 63(3), 2019, pp. 377–398 <https://doi.org/10.1007/s00168-019-00957-4>.

30 Cheng Ting-Fang, 'How a Small Taiwanese City Transformed the Global Chip Industry', *Nikkei Asia*, 15 December 2020 <https://asia.nikkei.com/Business/Technology/How-a-small-Taiwanese-city-transformed-the-global-chip-industry> [accessed 13 January 2021].

31 Jane Jacobs, *The Death and Life of Great American Cities* (New York: Random House, 1961), p. 31.

32 West, *Scale*, pp. 281–88.

33 'Bright Lights, Big Cities', *The Economist*, 4 February 2015 <https://www.economist.com/node/21642053> [accessed 20 March 2021].

34 Jeff Desjardins, 'By 2100 None of the World's Biggest Cities Will Be in China, the US or Europe', World Economic Forum, 20 July 2018 <https://www.weforum.org/agenda/2018/07/by-2100-none-of-the-worlds-biggest-cities-will-be-in-china-the-us-or-europe/> [accessed 20 March 2021].

35 'Cities Worldwide Will Struggle, but Will Avoid a Mass Exodus', *The Economist*, 17 November 2020 <https://www.economist.com/the-world-ahead/2020/11/17/cities-worldwide-will-struggle-but-will-avoid-a-mass-exodus> [accessed 20 March 2021].

36 'COVID-19 and the Myth of Urban Flight', *Knowledge@Wharton*, 1 December 2020 <https://knowledge.wharton.upenn.edu/article/covid-19-and-the-myth-of-urban-flight/> [accessed 13 January 2021].

37 Samrat Sharma, 'India's Rural-Urban Divide: Village Worker Earns Less than Half of City Peer', *Financial Express*, 12 December 2019 <https://www.financialexpress.com/economy/indias-rural-urban-divide-village-worker-earns-less-than-half-of-city-peer/1792245/> [accessed 18 March 2021].

38 'Is There Really an Ever-Widening Rural-Urban Divide in Europe', Euler Hermes Global, 17 July 2019 <https://www.eulerhermes.com/en_global/news-insights/economic-insights/Is-there-really-an-ever-widening-rural-urban-divide-in-Europe.html> [accessed 18 March 2021].

39 Patrick Greenfield, 'Uber Licence Withdrawal Disproportionate, Says Theresa May', *The Guardian*, 28 September 2017 <http://www.theguardian.com/technology/2017/sep/28/uber-licence-withdrawal-disproportionate-says-theresa-may> [accessed 23 March 2021].

40 'Why Cities and National Governments Clash over Migration', *Financial Times*, 4 June 2019 <https://www.ft.com/content/319ec1f6-5d25-11e9-840c-530737425559>.

41 John Perry Barlow, 'A Declaration of the Independence of Cyberspace', January 1996 <https://www.eff.org/cyberspace-independence> [accessed 7 January 2020].

42 Andrei Soldatov Borogan Irina, 'How the 1991 Soviet Internet Helped Stop a Coup and Spread a Message of Freedom', *Slate*, August 2016 <https://

slate.com/technology/2016/08/the-1991-soviet-internet-helped-stop-a-coup-and-spread-a-message-of-freedom.html> [accessed 31 July 2020].

43 Berhan Taye and Sage Cheung, 'The State of Internet Shutdowns in 2018', *Access Now*, 8 July 2019 <https://www.accessnow.org/the-state-of-internet-shutdowns-in-2018/> [accessed 19 July 2020].

44 Claudia Biancotti, 'India's Ill-Advised Pursuit of Data Localization', Pieterson Institute for International Economics, 9 December 2019 <https://www.piie.com/blogs/realtime-economic-issues-watch/indias-ill-advised-pursuit-data-localization> [accessed 20 March 2021].

45 DLA Piper, *Data Protection Laws of the World.*

46 Alan Beattie, 'Data Protectionism: The Growing Menace to Global Business', *Financial Times*, 14 May 2018 <https://medium.com/financial-times/data-protectionism-the-growing-menace-to-global-business-f994da37e9e2> [accessed 26 March 2021].

47 Ian Bremmer, 'Why We Need a World Data Organization. Now.', *GZERO*, 25 November 2019 <https://www.gzeromedia.com/why-we-need-a-world-data-organization-now> [accessed 20 March 2020].

48 Tanya Filer, Antonio and Weiss, 'Digital Minilaterals Are the Future of International Cooperation', *Brookings TechStream*, 16 October 2020 <https://www.brookings.edu/techstream/digital-minilaterals-are-the-future-of-international-cooperation/> [accessed 20 March 2021].

CHAPTER 7: THE NEW WORLD DISORDER

1 Damien McGuinness, 'How a Cyber Attack Transformed Estonia', *BBC News*, 27 April 2017 <https://www.bbc.com/news/39655415> [accessed 26 April 2021].

2 John Stuart Mill, *Principles of Political Economy* (1848), cited in Jong-Wha Lee and Ju Hyun Pyun, *Does Trade Integration Contribute to Peace?*, Asian Development Bank Working Paper (January 2009), p. 2.

3 J. W. Lee and J. H. Pyun, *Does Trade Integration Contribute to Peace,* Asian Development Bank Working Paper (January 2009), p. 18

4 Evan E. Hillebrand, 'Deglobalization Scenarios: Who Wins? Who Loses?', *Global Economy Journal*, vol. 10, issue 2 (2010).

5 Anouk Rigertink, *New Wars in Numbers: An Exploration of Various Datasets on Intra-State Violence*, MPRA Paper 45264 (University Library of Munich, 2012).

6 Uppsala Conflict Data Program, https://ucdp.uu.se/encyclopedia.

7 Azeem Azhar, 'AI and the Future of Warfare: My Conversation with General Sir Richard Barrons', *Exponential View*, 27 December 2019 <https://www.exponentialview.co/p/-ai-and-the-future-of-warfare> [accessed 18 September 2020].

8 Anshel Pfeffer, 'Zeev Raz Explains the Mind-Set of Israeli Pilots', *Jewish Chronicle*, 13 September 2012 <https://www.thejc.com/news/world/zeev-raz-explains-the-mind-set-of-israeli-pilots-1.36109> [accessed 9 September 2020].

9 John T. Correll, 'Air Strike at Osirak', *Air Force Magazine*, 1 April 2012 <https://www.airforcemag.com/article/0412osirak/> [accessed 26 April 2020].

10 Patrick Jackson, 'Osirak: Over the Reactor', *BBC News*, 5 June 2006 <http://news.bbc.co.uk/1/hi/world/middle_east/4774733.stm> [accessed 12 January 2021].

11 Ralph Langner, *To Kill a Centrifuge: A Technical Analysis of What Stuxnet's Creators Tried to Achieve* (The Langner Group, November 2013) <https://www.langner.com/to-kill-a-centrifuge/> [accessed 26 March 2020].

12 Ellen Nakashima and Joby Warrick, 'Stuxnet Was Work of U.S. and Israeli Experts, Officials Say', *Washington Post*, 2 June 2012 <https://www.washingtonpost.com/world/national-security/stuxnet-was-work-of-us-and-israeli-experts-officials-say/2012/06/01/gJQAlnEy6U_story.html> [accessed 26 March 2021].

13 Matthew Gooding, 'Cyber Attacks: Damaging Breaches Hit 96 per cent of UK Businesses Last Year', *Computer Business Review*, 5 August 2020 <https://www.cbronline.com/news/uk-businesses-cyber-attacks> [accessed 17 September 2020].

14 '37 Billion Data Records Leaked in 2020, a Growth of 140 per cent YOY – Atlas VPN' <https://atlasvpn.com/blog/37-billion-data-records-leaked-in-2020-a-growth-of-140-yoy> [accessed 24 February 2021].

15 Azeem Azhar, 'Cybersecurity in the Age of AI: My Conversation with Nicole Eagen', *Exponential View*, 27 December 2019 <https://www.exponentialview.co/p/cybersecurity-in-the-age-of-ai> [accessed 3 August 2020].

16 Matt Burgess, 'To Protect Putin, Russia Is Spoofing GPS Signals on a Massive Scale', *Wired*, 27 March 2019 <https://www.wired.co.uk/article/russia-gps-spoofing> [accessed 14 September 2020].

17 Andy Greenberg, 'The Untold Story of NotPetya, the Most Devastating Cyberattack in History', *Wired*, 22 August 2018 <https://www.wired.com/story/notpetya-cyberattack-ukraine-russia-code-crashed-the-world/> [accessed 3 August 2020].

18 Donghui Park, 'North Korea Cyber Attacks: A New Asymmetrical Military Strategy', Henry M. Jackson School of International Studies, 28 June 2016 <https://jsis.washington.edu/news/north-korea-cyber-attacks-new-asymmetrical-military-strategy/> [accessed 26 April 2021].

19 Edward White and Kang Buseong, 'Kim Jong Un's Cyber Army Raises Cash for North Korea', *Financial Times*, 17 June 2019 <https://www.ft.com/content/cbb28ab8-8ce9-11e9-a24d-b42f641eca37> [accessed 24 March 2021].

20 'North Korea GDP', Trading Economics <https://tradingeconomics.com/north-korea/gdp> [accessed 26 April 2021].

21 'North Korea GDP'.

22 Seth Jones, *Containing Tehran: Understanding Iran's Power and Exploiting Its Vulnerabilities* (Center for Strategic and International Studies, January 2020) <https://www.csis.org/analysis/containing-tehran-understanding-irans-power-and-exploiting-its-vulnerabilities> [accessed 21 June 2020].

23 'Six Russian GRU Officers Charged in Connection with Worldwide Deployment of Destructive Malware and Other Disruptive Actions in Cyberspace', US Department of Justice, 19 October 2020 <https://www.justice.gov/opa/pr/six-russian-gru-officers-charged-connection-worldwide-deployment-destructive-malware-and> [accessed 26 April 2021].

24 'MH17 Ukraine Plane Crash: What We Know', *BBC News*, 26 February 2020 <https://www.bbc.com/news/world-europe-28357880> [accessed 24 March 2021].

25 Andrew Marino, 'Sandworm Details the Group behind the Worst Cyberattacks in History', *The Verge*, 28 July 2020 <https://www.theverge.com/21344961/andy-greenberg-interview-book-sandworm-cyber-war-wired-vergecast> [accessed 14 September 2020].

26 Rae Ritchie, 'Maersk: Springing Back from a Catastrophic Cyber-Attack', I-CIO, August 2019 <https://www.i-cio.com/management/insight/item/maersk-springing-back-from-a-catastrophic-cyber-attack> [accessed 26 April 2021].

27 'The Sinkhole That Saved the Internet', *TechCrunch*, 8 July 2019 <https://social.techcrunch.com/2019/07/08/the-wannacry-sinkhole/> [accessed 12 January 2021].

28 'National Cyber Force Transforms Country's Cyber Capabilities to Protect the UK', GCHQ, 19 November 2020 <https://www.gchq.gov.uk/news/national-cyber-force> [accessed 2 January 2021].

29 Dustin Volz and Robert McMillan, 'Hack Suggests New Scope, Sophistication for Cyberattacks', *Wall Street Journal*, 18 December 2020 <https://www.wsj.com/articles/hack-suggests-new-scope-sophistication-for-cyberattacks-11608251360> [accessed 24 March 2021].

30 Simon Oxenham, '"I Was a Macedonian Fake News Writer"', *BBC Future*, 28 May 2019 <https://www.bbc.com/future/article/20190528-i-was-a-macedonian-fake-news-writer> [accessed 30 September 2020].

31 Soroush Vosoughi, Deb Roy and Sinan Aral, 'The Spread of True and False News Online', *Science*, 359(6380), 2018, pp. 1146–51 <https://doi.org/10.1126/science.aap9559>.

32 Sean Simpson, 'Fake News: A Global Epidemic Vast Majority (86 per cent) of Online Global Citizens Have Been Exposed to It', *Ipsos*, 11 June 2019 <https://www.ipsos.com/en-us/news-polls/cigi-fake-news-global-epidemic> [accessed 24 March 2021].

33 *Digital News Report 2018*, Reuters, p. 144.

34 Samantha Bradshaw and Philip N. Howard, *The Global Disinformation Order: 2019 Global Inventory of Organised Social Media Manipulation*, Working Paper 3 (Oxford, UK: Project on Computational Propaganda, 2019) <https://comprop.oii.ox.ac.uk/research/posts/the-global-disinformation-order-2019-global-inventory-of-organised-social-media-manipulation/#continue> [accessed 2 January 2021].

35 Diego Martin et al., *Trends in Online Influence Efforts* (Empirical Studies of Conflict Project, 2020) <https://esoc.princeton.edu/publications/trends-online-influence-efforts> [accessed 2 January 2021].

36 Gregory Winger, 'China's Disinformation Campaign in the Philippines', *The Diplomat*, 6 October 2020 <https://thediplomat.com/2020/10/chinas-disinformation-campaign-in-the-philippines/> [accessed 3 January 2021].

37 Jack Stubbs and Christopher Bing, 'Facebook, Twitter Dismantle Global Array of Disinformation Networks', *Reuters*, 8 October 2020 <https://www.reuters.com/article/cyber-disinformation-facebook-twitter-IDINKBN26T2XF> [accessed 24 March 2021].

38 Lorenzo Franceschi-Bicchierai, 'Russian Facebook Trolls Got Two Groups of People to Protest Each Other In Texas', *VICE*, 1 November 2017

<https://www.vice.com/en/article/3kvvz3/russian-facebook-trolls-got-people-to-protest-against-each-other-in-texas> [accessed 2 January 2021].

39 'How Covid-19 Is Revealing the Impact of Disinformation on Society', King's College London, 25 August 2020 <https://www.kcl.ac.uk/news/how-covid-19-is-revealing-the-impact-of-disinformation-on-society> [accessed 3 January 2021].

40 'Coronavirus: "Murder Threats" to Telecoms Engineers over 5G', *BBC News*, 23 April 2020 <https://www.bbc.com/news/newsbeat-52395771> [accessed 2 January 2021].

41 Wesley R. Moy and Kacper Gradon, 'COVID-19 Effects and Russian Disinformation Campaigns', *Homeland Security Affairs*, December 2020 <https://www.hsaj.org/articles/16533> [accessed 23 April 2021].

42 Simon Lewis, 'U.S. Says Russian-Backed Outlets Spread COVID-19 Vaccine "Disinformation"', *Reuters*, 8 March 2021 <https://www.reuters.com/article/us-usa-russia-covid-disinformation-IDUSKBN2B0016> [accessed 23 April 2021].

43 Peter Walker, 'New National Security Unit Set Up to Tackle Fake News in UK', *The Guardian*, 23 January 2018 <https://www.theguardian.com/politics/2018/jan/23/new-national-security-unit-will-tackle-spread-of-fake-news-in-uk>.

44 Brian Raymond, 'Forget Counterterrorism, the United States Needs a Counter-Disinformation Strategy', *Foreign Policy*, 15 October 2020 <https://foreignpolicy.com/2020/10/15/forget-counterterrorism-the-united-states-needs-a-counter-disinformation-strategy/> [accessed 24 March 2021].

45 Micah Zenko, 'Obama's Final Drone Strike Data', Council on Foreign Relations, 2017 <https://www.cfr.org/blog/obamas-final-drone-strike-data> [accessed 12 January 2021].

46 'Give and Take – Drone Technology Has Made Huge Strides', *The Economist*, 8 June 2017 <https://www.economist.com/technology-quarterly/2017/06/08/drone-technology-has-made-huge-strides> [accessed 14 October 2020].

47 'MQ-1B Predator', U.S. Air Force <https://www.af.mil/About-Us/Fact-Sheets/Display/Article/104469/mq-1b-predator/> [accessed 26 April 2021].

48 Reuters Staff, 'Factbox: The Global Hawk Drone Shot down by Iran', *Reuters*, 21 June 2019 <https://uk.reuters.com/article/us-mideast-iran-usa-factbox-IDUSKCN1TL29K> [accessed 14 October 2020].

49 'Bayraktar TB2 Unmanned Aerial Combat Vehicle', *Military Today* <http://military-today.com/aircraft/bayraktar_tb2.htm> [accessed 26 April 2021].

50 Umar Farooq, 'How Turkey Defied the U.S. and Became a Killer Drone Power', *The Intercept*, 14 May 2019 <https://theintercept.com/2019/05/14/turkey-second-drone-age/> [accessed 21 February 2021].

51 Associated Press, 'Major Saudi Arabia Oil Facilities Hit by Houthi Drone Strikes', *The Guardian*, 14 September 2019 <https://www.theguardian.com/world/2019/sep/14/major-saudi-arabia-oil-facilities-hit-by-drone-strikes> [accessed 26 April 2021].

52 Shaan Shaikh and Wes Rumbaugh, 'The Air and Missile War in Nagorno-Karabakh: Lessons for the Future of Strike and Defense', Center for Strategic and International Studies, 2020 <https://www.csis.org/analysis/air-and-missile-war-nagorno-karabakh-lessons-future-strike-and-defense> [accessed 2 January 2021].

53 Robyn Dixon, 'Azerbaijan's Drones Owned the Battlefield in Nagorno-Karabakh – and Showed Future of Warfare', *Washington Post*, 11 November2020<https://www.washingtonpost.com/world/europe/nagorno-karabkah-drones-azerbaijan-aremenia/2020/11/11/441bcbd2-193d-11eb-8bda-814ca56e138b_story.html> [accessed 12 January 2021].

54 Dan Sabbagh, 'UK Wants New Drones in Wake of Azerbaijan Military Success', *The Guardian*, 2020 <http://www.theguardian.com/world/2020/dec/29/uk-defence-secretary-hails-azerbaijans-use-of-drones-in-conflict> [accessed 12 January 2021].

55 Joseph Trevithick, 'China Conducts Test of Massive Suicide Drone Swarm Launched from a Box on a Truck', *The Drive*, 14 October 2020 <https://www.thedrive.com/the-war-zone/37062/china-conducts-test-of-massive-suicide-drone-swarm-launched-from-a-box-on-a-truck> [accessed 26 April 2021].

56 Joseph Trevithick, 'RAF Uses Autonomous Drone Swarm Loaded with Decoys to Overwhelm Mock Enemy Air Defenses', *The Drive*, 8 October 2020 <https://www.thedrive.com/the-war-zone/36950/raf-tests-swarm-loaded-with-britecloud-electronic-warfare-decoys-to-overwhelm-air-defenses> [accessed 26 April 2021].

57 Zachary Kallenborn and Philipp C. Bleek, 'Drones of Mass Destruction: Drone Swarms and the Future of Nuclear, Chemical, and Biological Weapons', *War on the Rocks*, 14 February 2019 <https://warontherocks.com/2019/02/drones-of-mass-destruction-drone-swarms-and-the-future-of-nuclear-chemical-and-biological-weapons/> [accessed 26 April 2021].

58 Missy Cummings, *The Human Role in Autonomous Weapon Design and Deployment*, 2014 <https://www.law.upenn.edu/live/files/3884-cummings-the-human-role-in-autonomous-weapons>.

59 Nick Statt, 'Skydio's AI-Powered Autonomous R1 Drone Follows You around in 4K', *The Verge*, 13 February 2018 <https://www.theverge.com/2018/2/13/17006010/skydio-r1-autonomous-drone-4k-video-recording-ai-computer-vision-mapping> [accessed 2 January 2021].

60 'Autonomous Weapons and the New Laws of War', *The Economist*, 19 January2019<https://www.economist.com/briefing/2019/01/19/autonomous-weapons-and-the-new-laws-of-war> [accessed 26 March 2021].

61 Burgess Laird, 'The Risks of Autonomous Weapons Systems for Crisis Stability and Conflict Escalation in Future U.S.-Russia Confrontations', Rand Corporation, 2020 <https://www.rand.org/blog/2020/06/the-risks-of-autonomous-weapons-systems-for-crisis.html> [accessed 2 January 2021].

62 Jindan-Karena Kann, 'Autonomous Weapons Systems and the Liability Gap, Part One: Introduction to Autonomous Weapons Systems and International Criminal Liability', *Rethinking SLIC*, 15 June 2019 <https://www.rethinkingslic.org/blog/criminal-law/51-autonomous-weapons-systems-and-the-liability-gap-part-one-introduction-to-autonomous-weapon-systems-and-international-criminal-liability> [accessed 26 March 2021].

63 Marta Bo, 'Autonomous Weapons and the Responsibility Gap in Light of the *Mens Rea* of the War Crime of Attacking Civilians in the ICC Statute in Weapons and Targeting', *Journal of International Criminal Justice*, March 2021, mqab005 <https://doi.org/10.1093/jicj/mqab005>.

64 Terri Moon Cronk, 'DOD's Cyber Strategy of the Past Year Outlined before Congress', US Department of Defense, 6 March 2020 <https://www.defense.gov/Explore/News/Article/Article/2103843/dods-cyber-strategy-of-past-year-outlined-before-congress/>.

65 Paul M. Nakasone and Michael Sulmeyer, 'How to Compete in Cyberspace', *Foreign Affairs*, 28 August 2020 <https://www.foreignaffairs.com/articles/united-states/2020-08-25/cybersecurity> [accessed 23 April 2021].

66 Azeem Azhar, 'AI and the Future of Warfare: A Conversation with General Sir Richard Barrons', *Exponential View*, 27 December 2019 <https://www.exponentialview.co/p/-ai-and-the-future-of-warfare> [accessed 18 September 2020].

67 Reid Standish, 'Why Is Finland Able to Fend Off Putin's Information War?', *Foreign Policy*, 1 March 2017 <https://foreignpolicy.com/2017/03/01/

why-is-finland-able-to-fend-off-putins-information-war/> [accessed 24 March 2021].

68 Eliza Mackintosh, 'Finland Is Winning the War on Fake News. Other Nations Want the Blueprint', CNN <https://edition.cnn.com/interactive/2019/05/europe/finland-fake-news-intl/> [accessed 24 March 2021].

69 Azeem Azhar, 'How Taiwan Is Using Technology to Foster Democracy', *Exponential View*, 23 October 2020 <https://www.exponentialview.co/p/how-taiwan-is-using-technology-to> [accessed 24 March 2021].

70 Ben Buchanan and Fiona S. Cunningham, 'Preparing the Cyber Battlefield: Assessing a Novel Escalation Risk in a Sino-American Crisis', *Texas National Security Review*, October 2020 <http://tnsr.org/2020/10/preparing-the-cyber-battlefield-assessing-a-novel-escalation-risk-in-a-sino-american-crisis/> [accessed 23 April 2021].

71 Jens Stoltenberg, 'Nato Will Defend Itself', *Prospect Magazine*, 27 August 2019 <https://www.prospectmagazine.co.uk/world/nato-will-defend-itself-summit-jens-stoltenberg-cyber-security> [accessed 12 March 2020]."

72 'CyberPeace Institute Calls for Accountability of Intrusion as a Service', CypérPeace Institute, 10 January 2021 <https://cyberpeaceinstitute.org/news/cyberpeace-institute-calls-for-accountability-of-intrusion-as-a-service>.

73 'Western Firms Should Not Sell Spyware to Tyrants', *The Economist*, 12 December 2019 <https://www.economist.com/leaders/2019/12/12/western-firms-should-not-sell-spyware-to-tyrants> [accessed 26 April 2021].

74 Bill Marczak et al., 'The Great IPwn: Journalists Hacked with Suspected NSO Group IMessage "Zero-Click" Exploit', The Citizen Lab, 20 December 2020 <https://citizenlab.ca/2020/12/the-great-ipwn-journalists-hacked-with-suspected-nso-group-imessage-zero-click-exploit/> [accessed 26 April 2021].

75 Vincen Boulanin, *Limits on Autonomy in Weapons Systems* (SIPRI, 2020) <https://www.sipri.org/sites/default/files/2020-06/2006_limits_of_autonomy.pdf>.

CHAPTER 8: EXPONENTIAL CITIZENS

1 'Dear Mark Zuckerberg. I shall not comply with your requirement to remove this picture.' <https://www.aftenposten.no/meninger/kommentar/i/G892Q/dear-mark-i-am-writing-this-to-inform-you-that-i-shall-not-comply-wit> [accessed 1 October 2020]."

2 https://www.bbc.co.uk/news/technology-37318031
3 'Facebook U-Turn over "Napalm Girl" Photograph', *BBC News*, 9 September 2016 <https://www.bbc.co.uk/news/technology-37318040> [accessed 1 October 2020].
4 Chris Hughes, 'It's Time to Break Up Facebook', *New York Times*, 9 May 2019<https://www.nytimes.com/2019/05/09/opinion/sunday/chris-hughes-facebook-zuckerberg.html> [accessed 4 April 2021].
5 Michael Sandel, *What Money Can't Buy: The Moral Limits of Markets* (London: Penguin Books, 2012).
6 'Data Never Sleeps 8.0', Domo <https://www.domo.com/learn/data-never-sleeps-8> [accessed 26 April 2021].
7 Lawrence Lessig, 'Code Is Law', *Harvard Magazine*, 1 January 2000 <https://harvardmagazine.com/2000/01/code-is-law-html> [accessed 2 October 2020].
8 'Apple and Google Partner on COVID-19 Contact Tracing Technology', Apple Newsroom, April 2020 <https://www.apple.com/uk/newsroom/2020/04/apple-and-google-partner-on-covid-19-contact-tracing-technology/> [accessed 13 October 2020].
9 Prabhash K. Dutta, '16 Lynchings in 2 Months. Is Social Media the New Serial Killer?', *India Today*, 2 July 2018 <https://www.indiatoday.in/india/story/16-lynchings-in-2-months-is-social-media-the-new-serial-killer-1275182-2018-07-02> [accessed 4 October 2020].
10 'India Lynchings: WhatsApp Sets New Rules after Mob Killings', *BBC News*, 20 July 2018 <https://www.bbc.co.uk/news/world-asia-india-44897714> [accessed 4 October 2020].
11 Reuters Staff, 'Facebook's WhatsApp Limits Users to Five Text Forwards to Curb Rumors', *Reuters*, 21 January 2019 <https://www.reuters.com/article/us-facebook-whatsapp-IDUSKCN1PF0TP> [accessed 4 October 2020].
12 Paul Mozur, 'A Genocide Incited on Facebook, With Posts From Myanmar's Military (Published 2018)', *New York Times*, 15 October 2018 <https://www.nytimes.com/2018/10/15/technology/myanmar-facebook-genocide.html> [accessed 8 January 2021].
13 Matthew Smith, 'Facebook Wanted to Be a Force for Good in Myanmar. Now It Is Rejecting a Request to Help With a Genocide Investigation', *Time*, 18 August 2020 <https://time.com/5880118/myanmar-rohingya-genocide-facebook-gambia/> [accessed 8 January 2021].

14 Madeleine Carlisle, 'Mark Zuckerberg Says Facebook's Decision to Not Take Down Kenosha Militia Page Was a Mistake', *Time*, 29 August 2020 <https://time.com/5884804/mark-zuckerberg-facebook-kenosha-shooting-jacob-blake/> [accessed 13 January 2021].

15 Nick Clegg, 'You and the Algorithm: It Takes Two to Tango', *Medium*, 31 March 2021 <https://nickclegg.medium.com/you-and-the-algorithm-it-takes-two-to-tango-7722b19aa1c2> [accessed 31 March 2021].

16 Carissa Véliz, 'Privacy Matters Because It Empowers Us All', *Aeon*, 2019 <https://aeon.co/essays/privacy-matters-because-it-empowers-us-all> [accessed 5 April 2021].

17 'Google Form S-1', 2004 <https://www.sec.gov/Archives/edgar/data/1288776/000119312504073639/ds1.htm> [accessed 5 April 2021].

18 Farhad Manjoo and Nadieh Bremer, 'I Visited 47 Sites. Hundreds of Trackers Followed Me', *New York Times*, 23 August 2019 <https://www.nytimes.com/interactive/2019/08/23/opinion/data-internet-privacy-tracking.html> [accessed 5 January 2021].

19 Brian X. Chen, 'I Downloaded the Information That Facebook Has on Me. Yikes.', *New York Times*, 11 April 2018 <https://www.nytimes.com/2018/04/11/technology/personaltech/i-downloaded-the-information-that-facebook-has-on-me-yikes.html> [accessed 4 April 2021].

20 Gabriel J. X. Dance, Nicholas Confessore and Michael LaForgia, 'Facebook Gave Device Makers Deep Access to Data on Users and Friends', *New York Times*, 3 June 2018 <https://www.nytimes.com/interactive/2018/06/03/technology/facebook-device-partners-users-friends-data.html, https://www.nytimes.com/interactive/2018/06/03/technology/facebook-device-partners-users-friends-data.html> [accessed 4 April 2021].

21 Megan Leonhardt, 'Democrats and Republicans in Congress Agree: The System That Determines Credit Scores Is "Broken"', CNBC, 27 February 2019 <https://www.cnbc.com/2019/02/27/american-consumer-credit-rating-system-is-broken.html> [accessed 26 April 2021].

22 Ellen Wright Clayton and others, 'The Law of Genetic Privacy: Applications, Implications, and Limitations', *Journal of Law and the Biosciences*, 6(1), 2019, pp. 1–36 <https://doi.org/10.1093/jlb/lsz007>.

23 Xinyuan Wang, 'Hundreds of Chinese Citizens Told Me What They Thought about the Controversial Social Credit System', *The Conversation*, 17 December 2019 <http://theconversation.com/hundreds-of-chinese-citizens-told-me-what-they-thought-about-the-controversial-social-credit-system-127467> [accessed 18 October 2020].

24 Azeem Azhar, 'Untangling China's Social Credit System', *Exponential View*, 6 February 2020 <https://www.exponentialview.co/p/-untangling-chinas-social-credit> [accessed 18 October 2020].

25 Heidi Ledford, 'Millions of Black People Affected by Racial Bias in Health-Care Algorithms', *Nature*, 574(7780), 2019, pp. 608–609 <https://doi.org/10.1038/d41586-019-03228-6>.

26 Jennifer McCoy and Murat Somer, 'Toward a Theory of Pernicious Polarization and How It Harms Democracies: Comparative Evidence and Possible Remedies', *The Annals of the American Academy of Political and Social Science*, 681(1), 2019, pp. 234–271 <https://doi.org/10.1177/0002716218818782>.

27 Paul Lazarsfeld and Robert Merton, 'Friendship as a Social Process: A Substantive and Methodological Analysis', in Morroe Berger, Theodore Abel, and Charles H. Page, eds, *Freedom and Control in Modern Society* (New York: Van Nostrand, 1954) <https://archive.org/stream/in.ernet.dli.2015.498862/2015.498862.Freedom-and_djvu.txt>.

28 Zeynep Tufekci, 'YouTube, the Great Radicalizer', *New York Times*, 10 March 2018 <https://www.nytimes.com/2018/03/10/opinion/sunday/youtube-politics-radical.html> [accessed 18 October 2020].

29 Tufekci, 'YouTube, the Great Radicalizer'.

30 Rino Ribeiro et al., 'Auditing Radicalization Pathways on YouTube', ArXiv: 1908.08313 [Cs], 2019 <http://arxiv.org/abs/1908.08313> [accessed 18 October 2020].

31 Gregory Waters and Robert Postings, *Spiders of the Caliphate: Mapping the Islamic State's Global Support Network on Facebook* (Counter Extremism Project, May 2018) <https://www.counterextremism.com/sites/default/files/Spiders per cent20of per cent20the per cent20Caliphate per cent20 per cent28May per cent202018 per cent29.pdf> [accessed 1 December 2020].

32 Avi Selk, '"Maybe Someone Dies": Facebook VP Justified Bullying, Terrorism as Costs of Network's "Growth"', *Washington Post*, 30 March 2018 <https://www.washingtonpost.com/news/the-switch/wp/2018/03/30/maybe-someone-dies-facebook-vp-justified-bullying-terrorism-as-costs-of-growth/> [accessed 4 April 2021].

33 Deepa Seetharaman and Jeff Horwitz, 'Facebook Executives Shut Down Efforts to Make the Site Less Divisive', *Wall Street Journal*, 26 May 2020 <https://www.wsj.com/articles/facebook-knows-it-encourages-division-top-executives-nixed-solutions-11590507499> [accessed 3 January 2021].

34 Karl Polanyi quoted in James Boyle, 'The Second Enclosure Movement and the Construction of the Public Domain', *Law and Contemporary Problems*, 66 (2003), 33–74 <https://doi.org/10.2139/ssrn.470983>.

35 Boyle, *The Second Enclosure*, p. 37.

36 James M. Acton, 'Cyber Warfare & Inadvertent Escalation', *Daedalus*, 149(2), 2020, pp. 133–149 <https://doi.org/10.1162/daed_a_01794>.

37 Schonfeld, Erick, 'The FriendFeedization of Facebook Continues: Bret Taylor Promoted to CTO', *TechCrunch*, 2 June 2010 <https://social.techcrunch.com/2010/06/02/facebook-cto-bret-taylor/> [accessed 5 April 2021].

38 Azeem Azhar, 'Fixing the Social Media Crisis: Sinan Aral & Azeem Azhar' [podcast]<https://podcasts.apple.com/gb/podcast/fixing-the-social-media-crisis/id1172218725?i=1000510469428> [accessed 5 April 2021].

39 will.i.am, 'We Need to Own Our Data as a Human Right—and Be Compensated for It', *The Economist*, 21 January 2019 <https://www.economist.com/open-future/2019/01/21/we-need-to-own-our-data-as-a-human-right-and-be-compensated-for-it> [accessed 18 October 2020].

40 Martin Tisné, 'It's Time for a Bill of Data Rights', *MIT Technology Review*, 14 December 2018 <https://www.technologyreview.com/2018/12/14/138615/its-time-for-a-bill-of-data-rights/> [accessed 8 October 2020].

41 Garrett Hardin, 'The Tragedy of the Commons', *Science*, 162(3859), 1968, pp. 1243–1248 <https://doi.org/10.1126/science.162.3859.1243>.

42 For a good discussion of Hardin and Ostrom's relative contributions to this debate, see Brett Frischmann, Alain Marciano and Giovanni Battista Ramello, 'Retrospectives: Tragedy of the Commons after 50 Years', *Journal of Economic Perspectives*, 33(4), 2019, pp. 211–228 <https://doi.org/10.1257/jep.33.4.211>.

43 Carol Rose, 'The Comedy of the Commons: Custom, Commerce, and Inherently Public Property', *The University of Chicago Law Review*, 53(3), 1986.

CONCLUSION: ABUNDANCE AND EQUITY

1 'Virological: Novel 2019 Coronavirus Genome', *Virological* <https://virological.org/> [accessed 10 January 2021].

2 Rino Rappuoli and others, 'Vaccinology in the Post−COVID-19 Era', *Proceedings of the National Academy of Sciences*, 118(3), 2021 <https://doi.org/10.1073/pnas.2020368118>.

3 '424B5', Moderna Stock Offering, SEC <https://www.sec.gov/Archives/edgar/data/1682852/000119312520033353/d871325d424b5.htm> [accessed 12 January 2021].
4 'COVID-19 Vaccine Tracker', March 2020 <https://www.raps.org/news-and-articles/news-articles/2020/3/covid-19-vaccine-tracker> [accessed 10 January 2021].
5 'Coronavirus: The World in Lockdown in Maps and Charts', *BBC News*, 6 April 2020 <https://www.bbc.com/news/world-52103747> [accessed 19 April 2021].
6 Hannah Mayer et al., 'AI Puts Moderna within Striking Distance of Beating COVID-19', *Harvard Business School Digital Initiative*, 24 November 2020 <https://digital.hbs.edu/artificial-intelligence-machine-learning/ai-puts-moderna-within-striking-distance-of-beating-covid-19/> [accessed 11 January 2021].
7 Carrie Arnold, 'How Computational Immunology Changed the Face of COVID-19 Vaccine Development', *Nature Medicine*, 15 July 2020 <https://doi.org/10.1038/d41591-020-00027-9>.
8 50 watt-hours is the equivalent of using a typical LED lightbulb for about six hours, using a modern iron for about two minutes, or running a laptop for about 30 minutes.
9 Niall P. Hanan and Julius Y. Anchang, 'Satellites Could Soon Map Every Tree on Earth', *Nature*, 14 October 2020 <https://doi.org/10.1038/d41586-020-02830-3>.
10 This idea was first espoused in 1865 by William Jevons in his essay 'The Coal Question'.
11 Sheila Jasanoff, *The Ethics of Invention* (New York: W. W. Norton and Company, 2016).
12 Melvin Kranzberg, 'Technology and History: "Kranzberg's Laws"', *Technology and Culture*, 27(3), 1986, pp. 544–560 <https://doi.org/10.2307/3105385>.
13 Jasanoff, *The Ethics of Invention*.

Select Bibliography

Acemoglu, Daron, Claire LeLarge and Pascual Restrepo, *Competing with Robots: Firm-Level Evidence from France*, Working Paper Series (National Bureau of Economic Research, February 2020) <https://doi.org/10.3386/w26738>

Acemoglu, Daron, and Pascual Restrepo, *Robots and Jobs: Evidence from US Labor Markets*, Working Paper Series (National Bureau of Economic Research, March 2017) <https://doi.org/10.3386/w23285>

Acton, James M., 'Cyber Warfare & Inadvertent Escalation', *Daedalus*, 149(2), 2020, pp. 133–49 <https://doi.org/10.1162/daed_a_01794>

Agrawal, Ravi, 'The Hidden Benefits of Uber', *Foreign Policy*, 16 July 2018 <https://foreignpolicy.com/2018/07/16/why-india-gives-uber-5-stars-gig-economy-jobs/> [accessed 21 September 2020]

Allen, Robert C., 'Engels' Pause: Technical Change, Capital Accumulation, and Inequality in the British Industrial Revolution', *Explorations in Economic History*, 46(4), 2009, pp. 418–35 <https://doi.org/10.1016/j.eeh.2009.04.004>

Almenberg, Johan, and Christer Gerdes, 'Exponential Growth Bias and Financial Literacy', *Applied Economics Letters*, 19(17), 2012, pp. 1693–96 <https://doi.org/10.1080/13504851.2011.652772>

Anthony, Scott D., 'Kodak's Downfall Wasn't About Technology', *Harvard Business Review*, 15 July 2016 <https://hbr.org/2016/07/kodaks-downfall-wasnt-about-technology> [accessed 14 December 2020]

'Arithmetic, Population and Energy – a Talk by Al Bartlett' <https://www.albartlett.org/presentations/arithmetic_population_energy.html> [accessed 3 December 2020]

Arnold, Carrie, 'How Computational Immunology Changed the Face of COVID-19 Vaccine Development', *Nature Medicine*, 15 July 2020 <https://doi.org/10.1038/d41591-020-00027-9>

Arthur, W. Brian, 'Increasing Returns and the New World of Business', *Harvard Business Review*, 1 July 1996 <https://hbr.org/1996/07/increasing-returns-and-the-new-world-of-business> [accessed 31 July 2020]

———, *The Nature of Technology: What It Is and How It Evolves* (London: Penguin Books, 2009)

Autor, David, David Dorn, Lawrence F. Katz, Christina Patterson, Chicago Booth and John Van Reenen, 'The Fall of the Labor Share and the Rise of Superstar Firms', *The Quarterly Journal of Economics*, 135(2), May 2020, pp. 645–709 <https://doi.org/10.1093/qje/qjaa004>

Banerjee, Ritwik, Joydeep Bhattacharya and Priyama Majumdar, 'Exponential-Growth Prediction Bias and Compliance with Safety Measures Related to COVID-19', *Social Science & Medicine*, 268, January 2021, 113473 <https://doi.org/10.1016/j.socscimed.2020.113473>

Barclay, Andrew, 'Why Is It So Hard to Make a Truly Self-Driving Car?', *South China Morning Post*, 5 July 2018 <https://www.scmp.com/abacus/tech/article/3028605/why-it-so-hard-make-truly-self-driving-car> [accessed 11 January 2021]

Benson, Christopher L., Giorgio Triulzi, and Christopher L. Magee, 'Is There a Moore's Law for 3D Printing?', *3D Printing and Additive Manufacturing*, 5(1), March 2018, pp. 53–62 <https://doi.org/10.1089/3dp.2017.0041>

Bessen, James, *Learning by Doing: The Real Connection between Innovation, Wages, and Wealth* (New Haven: Yale University Press, 2015)

Biancotti, Claudia, 'India's Ill-Advised Pursuit of Data Localization', Peterson Institute for International Economics, 9 December 2019 <https://www.piie.com/blogs/realtime-economic-issues-watch/indias-ill-advised-pursuit-data-localization> [accessed 20 March 2021]

Bo, Marta, 'Autonomous Weapons and the Responsibility Gap in Light of the *Mens Rea* of the War Crime of Attacking Civilians in the ICC Statute in Weapons and Targeting', *Journal of International Criminal Justice*, 2021, mqab005 <https://doi.org/10.1093/jicj/mqab005>

Bobier, Jean-François, Philipp Gerbert, Jens Burchardt, and Antoine Gourévich, 'A Quantum Advantage in Fighting Climate Change', BCG Global, 22 January 2020 <https://www.bcg.com/publications/2020/quantum-advantage-fighting-climate-change> [accessed 23 March 2021]

Boyer, Pascal, and Michael Bang Petersen, 'Folk-Economic Beliefs: An Evolutionary Cognitive Model', *Behavioral and Brain Sciences*, 41, 2018, E158 <https://doi.org/10.1017/S0140525X17001960>

Bradshaw, Samantha, and Philip N. Howard, *The Global Disinformation Order: 2019 Global Inventory of Organised Social Media Manipulation* (The Project on Computational Propaganda, 26 September 2019) <https://comprop.oii.ox.ac.uk/research/posts/the-global-disinformation-order-2019-global-inventory-of-organised-social-media-manipulation/#continue> [accessed 2 January 2021]

Bremmer, Ian, 'Why We Need a World Data Organization. Now.', *GZERO*, 25 November 2019 <https://www.gzeromedia.com/why-we-need-a-world-data-organization-now> [accessed 20 March 2020]

Brynjolfsson, Erik, and Andrew McAfee, *The Second Machine Age: Work, Progress, and Prosperity in a Time of Brilliant Technologies* (New York: W. W. Norton & Company, 2014)

Buchanan, Ben, and Fiona S. Cunningham, 'Preparing the Cyber Battlefield: Assessing a Novel Escalation Risk in a Sino-American Crisis', *Texas National Security Review*, October 2020 <http://tnsr.org/2020/10/preparing-the-cyber-battlefield-assessing-a-novel-escalation-risk-in-a-sino-american-crisis/> [accessed 23 April 2021]

Choudary, Sangeet Paul, *Platform Scale: How an Emerging Business Model Helps Startups Build Large Empires with Minimum Investment*, 1st edn (Singapore: Platform Thinking Labs Pte. Ltd, 2015)

Clark, Gregory, *A Farewell to Alms: A Brief Economic History of the World*, The Princeton Economic History of the Western World (Princeton, NJ: Princeton Univ. Press, 2007)

Clayton, Ellen Wright, Barbara J. Evans, James W. Hazel, and Mark A. Rothstein, 'The Law of Genetic Privacy: Applications, Implications, and Limitations', *Journal of Law and the Biosciences*, 6(1), 2019, pp. 1–36 <https://doi.org/10.1093/jlb/lsz007>

Clegg, Nick, 'You and the Algorithm: It Takes Two to Tango', *Medium*, 31 March 2021 <https://nickclegg.medium.com/you-and-the-algorithm-it-takes-two-to-tango-7722b19aa1c2> [accessed 31 March 2021]

Coase, R. H., 'The Nature of the Firm', *Economica*, 4(16), November 1937, pp. 386–405 <https://doi.org/10.1111/j.1468-0335.1937.tb00002.x>

Cortada, James W., *The Computer in the United States: From Laboratory to Market* (Armonk, NY: M. E. Sharpe, 1993)

Curmi, Elizabeth, Aakash Doshi, Gregory R. Badishkanian, Nick Coulter, David Driscoll, P.J. Juvekar and others, 'Feeding the Future', *Citi Global Perspectives and Solutions*, November 2018 <https://www.citivelocity.com/citigps/feeding-the-future/> [accessed 18 March 2021]

Daugherty, Paul R., H. James Wilson and Paul Michelman, 'Revisiting the Jobs That Artificial Intelligence Will Create', *MIT Sloan Management Review* (Summer2017) <https://sloanreview.mit.edu/article/revisiting-the-jobs-artificial-intelligence-will-create/>

Diamandis, Peter H., and Steven Kotler, *Abundance: The Future Is Better Than You Think* (New York: Free Press, 2014)

Dustman, Eric, Kareem Elwakil and Miguel Smart, *Metals 3D Printing: Closing the Cost Gap and Getting to Value* (PricewaterhouseCoopers, 2019) <https://www.strategyand.pwc.com/gx/en/insights/2019/metals-3D-printing.html>

Engels, Frederick, *The Conditions of the Working-Class in England Index* (Panther, 1969)<https://www.marxists.org/archive/marx/works/1845/condition-working-class/> [accessed 18 December 2020]

Florida, Richard, 'The World Is Spiky', *Atlantic Monthly*, October 2005, pp. 48–51

'Frederick Winslow Taylor: Father of Scientific Management Thinker', The British Library <https://www.bl.uk/people/frederick-winslow-taylor> [accessed 29 March 2021]

Frey, Carl, and Robert Atkinson, 'Will AI Destroy More Jobs Than It Creates Over the Next Decade?', *Wall Street Journal*, 1 April 2019 <https://www.wsj.com/articles/will-ai-destroy-more-jobs-than-it-creates-over-the-next-decade-11554156299> [accessed 11 January 2021]

Frey, Carl Benedikt, *The Technology Trap: Capital, Labor, and Power in the Age of Automation* (Princeton, NJ: Princeton University Press, 2019)

Frey, Carl Benedikt, and Michael Osborne, 'The Future of Employment: How Susceptible Are Jobs to Computerisation?', Oxford Martin School working paper, 17 September 2013 <https://www.oxfordmartin.ox.ac.uk/publications/the-future-of-employment/> [accessed 14 September 2020]

Friedman, Milton, 'A Friedman Doctrine – The Social Responsibility of Business Is to Increase Its Profits', *New York Times*, 13 September 1970 <https://www.nytimes.com/1970/09/13/archives/a-friedman-doctrine-the-social-responsibility-of-business-is-to.html> [accessed 13 July 2020]

Frischmann, Brett, Alain Marciano and Giovanni Battista Ramello, 'Retrospectives: Tragedy of the Commons after 50 Years', *Journal of Economic Perspectives* 33(4), 2019, pp. 211–28 <https://doi.org/10.1257/jep.33.4.211>

Giddens, Anthony, *The Constitution of Society: Outline of the Theory of Structuration* (Berkeley, CA: University of California Press, 1986)

Giuliano, Genevieve, Sanggyun Kang, and Quan Yuan, 'Agglomeration Economies and Evolving Urban Form', *The Annals of Regional Science*, 63(3), 2019, pp. 377–398 <https://doi.org/10.1007/s00168-019-00957-4>

Gofman, Michael, and Zhao Jin, 'Artificial Intelligence, Human Capital, and Innovation', *SSRN Electronic Journal*, October 2019 <https://doi.org/10.2139/ssrn.3449440>

Greenberg, Andy, 'The Untold Story of NotPetya, the Most Devastating Cyberattack in History', *Wired*, 22 August 2018 <https://www.wired.com/story/notpetya-cyberattack-ukraine-russia-code-crashed-the-world/> [accessed 3 August 2020]

Grim, Cheryl, 'What Drives Productivity Growth?', United States Census Bureau, 2019 <https://www.census.gov/library/stories/2019/10/what-drives-productivity-growth.html> [accessed 2 September 2020]

Hanan, Niall P., and Julius Y. Anchang, 'Satellites Could Soon Map Every Tree on Earth', *Nature*, 2020 <https://doi.org/10.1038/d41586-020-02830-3>

Hannes, T. X., 'Will Technological Convergence Reverse Globalization?', National Defense University Press, 12 July 2016 <http://ndupress.ndu.edu/Media/News/News-Article-View/Article/834357/will-technological-convergence-reverse-globalization/> [accessed 6 September 2020]

Hansen, Espen Egil, 'Dear Mark Zuckerberg. I shall not comply with your requirement to remove this picture.', *Aftenposten*, 8 September 2016 <https://www.aftenposten.no/meninger/kommentar/i/G892Q/dear-mark-i-am-writing-this-to-inform-you-that-i-shall-not-comply-wit> [accessed 1 October 2020]

Hardin, Garrett, 'The Tragedy of the Commons', *Science*, 162(3859), 1968, pp. 1243–1248 <https://doi.org/10.1126/science.162.3859.1243>

Haskel, Jonathan, and Stian Westlake, *Capitalism without Capital: The Rise of the Intangible Economy* (Princeton, NJ: Princeton University Press, 2018)

Holzhausen, Arne, and Timo Wochner, 'Is There Really an Ever-Widening Rural-Urban Divide in Europe', Euler Hermes Global, 17 July 2019 <https://www.eulerhermes.com/en_global/news-insights/economic-insights/Is-there-really-an-ever-widening-rural-urban-divide-in-Europe.html> [accessed 18 March 2021]

'How Covid-19 Is Revealing the Impact of Disinformation on Society', King's College London, 25 August 2020 <https://www.kcl.ac.uk/news/how-covid-19-is-revealing-the-impact-of-disinformation-on-society> [accessed 3 January 2021]

Howe, Jeff, 'The Rise of Crowdsourcing', *Wired*, 1 June 2006, <https://www.wired.com/2006/06/crowds/> [accessed 28 September 2020]

Intangible Asset Market Value Study, Ocean Tomo <https://www.oceantomo.com/intangible-asset-market-value-study/> [accessed 27 August 2020]

Jacobs, Jane, *The Death and Life of Great American Cities* (New York: Random House, 1961)

Janeway, William H., *Doing Capitalism in the Innovation Economy: Reconfiguring the Three-Player Game between Markets, Speculators and the State* (Cambridge, UK: Cambridge University Press, 2018)

Jean, Joel, Patrick Richard Brown and Institute of Physics (Great Britain), *Emerging Photovoltaic Technologies* (Bristol, UK: IOP Publishing, 2020) <https://iopscience.iop.org/book/978-0-7503-2152-5> [accessed 12 October 2020]

Kallenborn, Zachary, and Philipp Bleek, 'Drones of Mass Destruction: Drone Swarms and the Future of Nuclear, Chemical, and Biological Weapons', *War On the Rocks*, 14 February 2019 <https://warontherocks.com/2019/02/drones-of-mass-destruction-drone-swarms-and-the-future-of-nuclear-chemical-and-biological-weapons/> [accessed 26 April 2021]

Karlis, Nicole, 'DoorDash Drivers Make an Average of $1.45 an Hour, Analysis Finds', *Salon*, 19 January 2020 <https://www.salon.com/2020/01/19/doordash-drivers-make-an-average-of-145-an-hour-analysis-finds/> [accessed 27 March 2021]

Katsifis, Dimitrios, 'The CMA Publishes Final Report on Online Platforms and Digital Advertising', *The Platform Law Blog*, 6 July 2020 <https://theplatformlaw.blog/2020/07/06/the-cma-publishes-final-report-on-online-platforms-and-digital-advertising/> [accessed 30 March 2021]

———, 'Economic Possibilities for Our Grandchildren', in *Essays in Persuasion* (London: Palgrave Macmillan UK, 2010), pp. 321–332 <https://doi.org/10.1007/978-1-349-59072-8_25>

Khan, Lina M., 'Amazon's Antitrust Paradox', *Yale Law Journal*, 126(3), January 2017, pp. 710–805

Klenert, David, Enrique Fernández-Macías and José-Ignacio Antón, 'Don't Blame It on the Machines: Robots and Employment in Europe', *VoxEU*, 24 February 2020 <https://voxeu.org/article/dont-blame-it-machines-robots-and-employment-europe> [accessed 10 September 2020]

Klinger, Joel, Juan Mateos-Garcia, and Konstantinos Stathoulopoulos, 'A Narrowing of AI Research?', arXiv:2009.10385 [cs.CY], 2020 <http://arxiv.org/abs/2009.10385> [accessed 30 March 2021]

Kranzberg, Melvin, 'Technology and History: "Kranzberg's Laws"', *Technology and Culture*, 27(3), 1986, pp. 544–560 <https://doi.org/10.2307/3105385>

Kurzweil, Ray, *The Age of Spiritual Machines: When Computers Exceed Human Intelligence* (New York, NY: Penguin, 2000)

———, 'The Law of Accelerating Returns', Kurzweil.net, 2001 <https://www.kurzweilai.net/the-law-of-accelerating-returns> [accessed 29 July 2020]

———, 'Average Transistor Price', Singularity.com <http://www.singularity.com/charts/page59.html> [accessed 10 March 2021]

Laird, Burgess, 'The Risks of Autonomous Weapons Systems for Crisis Stability and Conflict Escalation in Future U.S.-Russia Confrontations', Rand Corporation, 2020 <https://www.rand.org/blog/2020/06/the-risks-of-autonomous-weapons-systems-for-crisis.html> [accessed 2 January 2021]

Langner, Ralph, *To Kill a Centrifuge: A Technical Analysis of What Stuxnet's Creators Tried to Achieve* (The Langner Group, November 2013) <https://www.langner.com/to-kill-a-centrifuge/> [accessed 26 March 2020]

Lazarsfeld, Paul F., and Robert K. Merton, 'Friendship as a Social Process: A Substantive and Methodological Analysis', in Morroe Berger, Theodore Abel, and Charles H. Page, eds, *Freedom and Control in Modern Society*, (New York: Van Nostrand, 1954), pp. 18–66 <https://archive.org/stream/in.ernet.dli.2015.498862/2015.498862.Freedom-and_djvu.txt>

Ledford, Heidi, 'Millions of Black People Affected by Racial Bias in Health-Care Algorithms', *Nature*, 574(7780), 2019, pp. 608–609 <https://doi.org/10.1038/d41586-019-03228-6>

Lee, Neil, and Stephen Clarke, 'Do Low-Skilled Workers Gain from High-Tech Employment Growth? High-Technology Multipliers, Employment and Wages in Britain', *Research Policy*, 48(9), November 2019, 103803 <https://doi.org/10.1016/j.respol.2019.05.012>

Leiserson, Charles E., Neil C. Thompson, Joel S. Emer, Bradley C. Kuszmaul, Butler W. Lampson, Daniel Sanchez et al., 'There's Plenty of Room at the Top: What Will Drive Computer Performance after Moore's Law?', *Science*, 368(6495), June 2020 <https://doi.org/10.1126/science.aam9744>

Lessig, Lawrence, 'Code Is Law', *Harvard Magazine*, 1 January 2000 <https://harvardmagazine.com/2000/01/code-is-law-html> [accessed 2 October 2020]

Levinson, Marc, *Outside the Box: How Globalization Changed from Moving Stuff to Spreading Ideas* (Princeton, NJ: Princeton University Press, 2020)

Levy, Matthew R., and Joshua Tasoff, 'Exponential-Growth Bias and Overconfidence', *Journal of Economic Psychology*, 58, 2017, pp. 1–14 <https://doi.org/10.1016/j.joep.2016.11.001>

Maddison, Angus, *Contours of the World Economy, 1–2030 AD: Essays in Macro-Economic History* (Oxford; New York: Oxford University Press, 2007)

Mann, Jindan-Karena, 'Autonomous Weapons Systems and the Liability Gap, Part One: Introduction to Autonomous Weapons Systems and International Criminal Liability', *Rethinking SLIC*, 2019 <https://www.rethinkingslic.org/blog/criminal-law/51-autonomous-weapons-systems-and-the-liability-gap-part-one-introduction-to-autonomous-weapon-systems-and-international-criminal-liability> [accessed 26 March 2021]

Manyika, James, Susan Lund, Kelsey Robinson, John Valentino and Richard Dobbs, *Connecting Talent with Opportunity in the Digital Age* (McKinsey & Company, 1 June 2015) <https://www.mckinsey.com/featured-insights/employment-and-growth/connecting-talent-with-opportunity-in-the-digital-age> [accessed 6 October 2020]

Manyika, James, Jan Michke, Jacques Bughin, Jonathan Woetzel, Mekala Krishnan and Samuel Cudre, *A New Look at the Declining Labor Share of Income in the United States* (McKinsey Global Institute, May 2019) <https://www.mckinsey.com/featured-insights/employment-and-growth/a-new-look-at-the-declining-labor-share-of-income-in-the-united-states>

Manyika, James, Sree Ramaswamy, Jacques Bughin, Jonathan Woetzel, Michael Birshan and Zubin Nagpal, *'Superstars': The Dynamics of Firms, Sectors, and Cities Leading the Global Economy* (McKinsey & Company, 24 October 2018) <https://www.mckinsey.com/featured-insights/innovation-and-growth/superstars-the-dynamics-of-firms-sectors-and-cities-leading-the-global-economy> [accessed 19 December 2020]

Martin, Diego A., Jacob N. Shapiro and Julia Ilhardt, *Trends in Online Influence Efforts* (Empirical Studies of Conflict Project, 2020) <https://esoc.princeton.edu/publications/trends-online-influence-efforts> [accessed 2 January 2021]

Mayer, Hannah, Karim Lakhani, Marco Iansiti and Kerry Herman, 'AI Puts Moderna within Striking Distance of Beating COVID-19', *Harvard Business School Digital Initiative*, 24 November 2020 <https://digital.hbs.edu/artificial-intelligence-machine-learning/ai-puts-moderna-within-striking-distance-of-beating-covid-19/> [accessed 11 January 2021]

Mazzucato, Mariana, *The Entrepreneurial State: Debunking Public vs. Private Sector Myths* (London: Penguin Books, 2018)

McCoy, Jennifer, and Murat Somer, 'Toward a Theory of Pernicious Polarization and How It Harms Democracies: Comparative Evidence and Possible Remedies', *The Annals of the American Academy of Political and Social Science*, 681(1), 2019, pp. 234–271 <https://doi.org/10.1177/0002716218818782>

McDonald, Duff, *The Firm: The Story of McKinsey and Its Secret Influence on American Business* (New York: Simon and Schuster, 2014)

Meadows, Donella H., and Club of Rome, eds., *The Limits to Growth: A Report for the Club* (New York: Potomac Associates, 1972)

Mody, Cyrus C. M., *The Long Arm of Moore's Law: Microelectronics and American Science*, Inside Technology (Cambridge, MA: The MIT Press, 2017)

Molina, Gustavo Marcelo, and Pedro Enrique Mercado, 'Modelling and Control Design of Pitch-Controlled Variable Speed Wind Turbines', in Ibrahim H. Al-Bahadly, ed., *Wind Turbines* (InTech, 2011) <https://doi.org/10.5772/15880>

Moore, G. E., 'Cramming More Components onto Integrated Circuits', *Proceedings of the IEEE*, 86(1), 1965, pp. 82–85 <https://doi.org/10.1109/JPROC.1998.658762>

Moore, Jon, and Nat Bullard, *BNEF Executive Factbook 2021* (BloombergNEF, 2 March 2021)

Moravec, Hans P., *Mind Children: The Future of Robot and Human Intelligence* (Cambridge, MA: Harvard University Press, 1988)

Moy, Wesley, and Kacper Gradon, 'COVID-19 Effects and Russian Disinformation Campaigns', *Homeland Security Affairs*, 16, article 8, December 2020 <https://www.hsaj.org/articles/16533> [accessed 23 April 2021]

Mueller, Milton, *Will the Internet Fragment? Sovereignty, Globalization, and Cyberspace*, Digital Futures (Cambridge, UK: Polity Press, 2017)

Naam, Ramez, in conversation with Azeem Azhar, 'The Exponential March of Solar Energy', *Exponential View*, 14 May 2020 <https://www.exponentialview.co/p/-the-exponential-march-of-solar-energy> [accessed 1 August 2020]

Nagy, Béla, J. Doyne Farmer, Quan M. Bui and Jessika E. Trancik, 'Statistical Basis for Predicting Technological Progress', *PLOS ONE*, 8(2), 2013, e52669 <https://doi.org/10.1371/journal.pone.0052669>

Nakasone, Paul M., and Michael Sulmeyer, 'How to Compete in Cyberspace', *Foreign Affairs*, 2021 <https://www.foreignaffairs.com/articles/united-states/2020-08-25/cybersecurity> [accessed 23 April 2021]

Nelson, Richard R., *Technology, Institutions, and Economic Growth* (Cambridge, MA: Harvard University Press, 2005)

Nemet, Gregory F., *How Solar Energy Became Cheap: A Model for Low-Carbon Innovation* (London: Routledge, 2019)

North, Douglass C., *Institutions, Institutional Change and Economic Performance* (Cambridge, UK: Cambridge University Press, 1990) <https://doi.org/10.1017/CBO9780511808678>

Ogburn, William Fielding, *Social Change With Respect to Culture and Original Nature* (New York: B. W. Huebsch, 1923)

O'Reilly, Tim, 'Network Effects in Data', *O'Reilly Radar*, 27 October 2008 <http://radar.oreilly.com/2008/10/network-effects-in-data.html> [accessed 9 December 2020]

Park, Donghui, 'North Korea Cyber Attacks: A New Asymmetrical Military Strategy', Henry M. Jackson School of International Studies, 28 June 2016 <https://jsis.washington.edu/news/north-korea-cyber-attacks-new-asymmetrical-military-strategy/> [accessed 26 April 2021]

Perez, Carlota, *Technological Revolutions and Financial Capital: The Dynamics of Bubbles and Golden Ages* (Cheltenham: Edward Elgar, 2003)

Polanyi, Michael, and Amartya Sen, *The Tacit Dimension* (Chicago, IL: University of Chicago Press, 2009)

Porter, Michael E., *Competitive Advantage: Creating and Sustaining Superior Performance* (New York: Free Press, 1998) <http://catalog.hathitrust.org/api/volumes/oclc/38281769.html> [accessed 21 March 2021]

Prahalad, C. K., and Gary Hamel, 'The Core Competence of the Corporation', *Harvard Business Review*, 1 May 1990 <https://hbr.org/1990/05/the-core-competence-of-the-corporation> [accessed 27 August 2020]

Rappuoli, Rino, Ennio De Gregorio, Giuseppe Del Giudice, Sanjay Phogat, Simone Pecetta, Mariagrazia Pizza et al., 'Vaccinology in the Post−COVID-19 Era', *Proceedings of the National Academy of Sciences*, 118(3), 2021, e2020368118 <https://doi.org/10.1073/pnas.2020368118>

Raymond, Brian, 'Forget Counterterrorism, the United States Needs a Counter-Disinformation Strategy', *Foreign Policy*, 15 October 2020 <https://foreignpolicy.com/2020/10/15/forget-counterterrorism-the-united-states-needs-a-counter-disinformation-strategy/> [accessed 24 March 2021]

Reillier, Laure Claire, and Benoit Reillier, *Platform Strategy: How to Unlock the Power of Communities and Networks to Grow Your Business* (London: Routledge, 2017)

Ribeiro, Manoel Horta, Raphael Ottoni, Robert West, Virgílio A. F. Almeida, and Wagner Meira, 'Auditing Radicalization Pathways on YouTube', ArXiv:1908.08313 [Cs], 2019 <http://arxiv.org/abs/1908.08313> [accessed 18 October 2020]

Rigertink, Anouk S., *New Wars in Numbers. An Exploration of Various Datasets on Intra-State Violence*, MPRA Paper 45264 (University Library of Munich, 2012) <https://mpra.ub.uni-muenchen.de/45264/>

Rose, Carol, 'The Comedy of the Commons: Custom, Commerce, and Inherently Public Property', *The Unviersity of Chicago Law Review*, 53(3), 1986, pp. 711–781

Rosenblat, Alex, Tamara Kneese and danah boyd, *Workplace Surveillance* (Data & Society Research Institute, 4 January 2017) <https://doi.org/10.31219/osf.io/7ryk4>

Rotman, David, 'We're Not Prepared for the End of Moore's Law', *MIT Technology Review*, 24 February 2020 <https://www.technologyreview.com/2020/02/24/905789/were-not-prepared-for-the-end-of-moores-law/> [accessed 11 March 2021]

Saval, Nikil, *Cubed: A Secret History of the Workplace* (New York: Anchor Books, 2015)

Sanger, David E., *The Perfect Weapon: War, Sabotage, and Fear in the Cyber Age* (New York: Crown Publishers, 2018)

Schmidt, Eric, and Jared Cohen, *The New Digital Age: Reshaping the Future of People, Nations and Business* (New York: Alfred A. Knopf, 2013)

Schonfeld, Erick, 'The FriendFeedization of Facebook Continues: Bret Taylor Promoted To CTO', *TechCrunch*, 2 June 2010 <https://social.techcrunch.com/2010/06/02/facebook-cto-bret-taylor/> [accessed 5 April 2021]

Shaikh, Shaan, and Wes Rumbaugh, 'The Air and Missile War in Nagorno-Karabakh: Lessons for the Future of Strike and Defense', Center for Strategic and International Studies, 2020 <https://www.csis.org/analysis/air-and-missile-war-nagorno-karabakh-lessons-future-strike-and-defense> [accessed 2 January 2021]

Smil, Vaclav, *Creating the Twentieth Century: Technical Innovations of 1867–1914 and Their Lasting Impact* (New York: Oxford University Press, 2005)

———, *Energy and Civilization: A History* (Cambridge, MA: MIT Press, 2017)

———, *Prime Movers of Globalization: The History and Impact of Diesel Engines and Gas Turbines* (Cambridge, MA: MIT Press, 2010)

Smith, Brad, and Carol Ann Browne, *Tools and Weapons: The Promise and the Peril of the Digital Age* (New York: Penguin Press, 2019)

Solé, Ricard V., *Phase Transitions*, Primers in Complex Systems (Princeton, NJ: Princeton University Press, 2011)

Standish, Reid, 'Why Is Finland Able to Fend Off Putin's Information War?', *Foreign Policy*, 1 March 2017 <https://foreignpolicy.com/2017/03/01/why-is-finland-able-to-fend-off-putins-information-war/> [accessed 24 March 2021]

Stango, Victor, and Jonathan Zinman, 'Exponential Growth Bias and Household Finance', *The Journal of Finance*, 64(6), December 2009, pp. 2807–49 <https://doi.org/10.1111/j.1540-6261.2009.01518.x>

Stephens, Zachary D., Skylar Y. Lee, Faraz Faghri, Roy H. Campbell, Chengxiang Zhai, Miles J. Efron et al., 'Big Data: Astronomical or Genomical?', *PLoS Biology*, 13(7), July 2015 <https://doi.org/10.1371/journal.pbio.1002195>

Stokel-Walker, Chris, 'Politicians Still Don't Understand the Tech Companies They're Supposed to Keep in Check. That's a Problem.', *Business Insider*, 10 October 2020 <https://www.businessinsider.com/tiktok-ban-hearings-politicians-senators-know-nothing-about-tech-2020-10> [accessed 12 April 2021]

Stoltenberg, Jens, 'Nato Will Defend Itself', *Prospect Magazine*, 27 August 2019 <https://www.prospectmagazine.co.uk/world/nato-will-defend-itself-summit-jens-stoltenberg-cyber-security> [accessed 12 March 2020]

Tarnoff, Ben, 'The Making of the Tech Worker Movement', *Logic Magazine*, 4 May 2020 <https://logicmag.io/the-making-of-the-tech-worker-movement/full-text/> [accessed 3 April 2021]

Thelen, Kathleen, *How Institutions Evolve: The Political Economy of Skills in Germany, Britain, the United States, and Japan* (Cambridge, UK: Cambridge University Press, 2004) <https://doi.org/10.1017/CBO9780511790997>

Thiel, Peter, 'Competition Is for Losers', *Wall Street Journal*, 12 September 2014 <https://online.wsj.com/articles/peter-thiel-competition-is-for-losers-1410535536> [accessed 9 October 2020]

Thurner, Stefan, Peter Klimek and Rudolf Hanel, *Introduction to the Theory of Complex Systems* (Oxford: Oxford University Press, 2018) <https://www.oxfordscholarship.com/view/10.1093/oso/9780198821939.001.0001/oso-9780198821939> [accessed 19 July 2020]

Tisné, Martin, 'It's Time for a Bill of Data Rights', *MIT Technology Review*, 14 December 2018 <https://www.technologyreview.com/2018/12/14/138615/its-time-for-a-bill-of-data-rights/> [accessed 8 October 2020]

Tufekci, Zeynep, 'YouTube, the Great Radicalizer', *New York Times*, 10 March2018<https://www.nytimes.com/2018/03/10/opinion/sunday/youtube-politics-radical.html> [accessed 18 October 2020]

Turck, Matt 'The Power of Data Network Effects', *Matt Turck* [blog], 2016 <https://mattturck.com/the-power-of-data-network-effects/> [accessed 3 August 2020]

UBS, *The Food Revolution*, July 2019 <https://www.ubs.com/global/en/ubs-society/our-stories/2019/future-of-food/_jcr_content/mainpar/toplevelgrid_1749059381/col1/linklist/link.1695495471.file/bGluay9wYXRoPS9jb250ZW50L2RhbS91YnMvZ2xvYmFsL3Vicy1zb2NpZXR5LzIwMTkvZm9vZC1yZXZvbHV0aW9uLWp1bHkucGRm/food-revolution-july.pdf>

Van Reenen, John, and Christina Patterson, 'Research: The Rise of Superstar Firms Has Been Better for Investors than for Employees', *Harvard Business Review*, 11 May 2017 <https://hbr.org/2017/05/research-the-rise-of-superstar-firms-has-been-better-for-investors-than-for-employees> [accessed 2 September 2020]

Van Reenen, John, Christina Patterson, Lawrence Katz, David Dorn and David Autor, 'The Fall of the Labor Share and the Rise of Superstar Firms', CEP Discussion Paper No. 1482, October 2019 <http://cep.lse.ac.uk/pubs/download/dp1482.pdf>

van Zanden, J. L., *The Long Road to the Industrial Revolution: The European Economy in a Global Perspective, 1000–1800*, Global Economic History Series (Leiden; Boston: Brill, 2009)

Véliz, Carissa, 'Privacy Matters Because It Empowers Us All', *Aeon*, 2019 <https://aeon.co/essays/privacy-matters-because-it-empowers-us-all> [accessed 5 April 2021]

Vosoughi, Soroush, Deb Roy and Sinan Aral, 'The Spread of True and False News Online', *Science*, 359(6380), 2018, pp. 1146–51 <https://doi.org/10.1126/science.aap9559>

Voth, Hans-Joachim, 'The Longest Years: New Estimates of Labor Input in England, 1760–1830', *Journal of Economic History*, 61(4), December 2001, pp. 1065–82 <https://doi.org/10.1017/S0022050701042085>

Wagenaar, William A., and Sabato D. Sagaria, 'Misperception of Exponential Growth', *Perception & Psychophysics*, 18(6), November 1975, pp. 416–22 <https://doi.org/10.3758/BF03204114>

Waters, Gregory and Robert Postings, *Spiders of the Caliphate: Mapping the Islamic State's Global Support Network on Facebook* (Counter Extremism Project, May 2018) <https://www.counterextremism.com/sites/default/files/Spiders%20of%20the%20Caliphate%20%28May%202018%29.pdf> [accessed 1 December 2020]

Weiss, Antonio and Tanya Filer, 'Digital Minilaterals Are the Future of International Cooperation', *Brookings TechStream*, 16 October 2020 <https://www.brookings.edu/techstream/digital-minilaterals-are-the-future-of-international-cooperation/> [accessed 20 March 2021]

West, Geoffrey B., *Scale: The Universal Laws of Growth, Innovation, Sustainability, and the Pace of Life in Organisms, Cities, Economies, and Companies* (New York: Penguin Press, 2017)

Westlake, Stian, quoted in Azeem Azhar, 'Understanding the Intangible Economy', *Exponential View*, 5 July 2019 <https://www.exponentialview.co/p/capitalism-without-capital> [accessed 28 August 2020]

'Why Has The Cost Of Genome Sequencing Declined So Rapidly?' Biostars forum <https://www.biostars.org/p/42753/> [accessed 28 July 2020]

Willcocks, Leslie, 'Robo-Apocalypse Cancelled? Reframing the Automation and Future of Work Debate', *Journal of Information Technology*, 35(4), 2020, pp. 286–302 <https://doi.org/10.1177/0268396220925830>

will.i.am, 'We Need to Own Our Data as a Human Right – and Be Compensated for It', *The Economist*, 21 January 2019 <https://www.economist.com/open-future/2019/01/21/we-need-to-own-our-data-as-a-human-right-and-be-compensated-for-it> [accessed 18 October 2020]

Wright, T. P., 'Factors Affecting the Cost of Airplanes', *Journal of the Aeronautical Sciences*, 3(4), February 1936, p. 122–128 <https://doi.org/10.2514/8.155>

Yeo, Lijin, 'The U.S. Rideshare Industry: Uber vs. Lyft', *Bloomberg Second Measure*, 2020 <https://secondmeasure.com/datapoints/rideshare-industry-overview/> [accessed 23 September 2020]

Youn, Hyejin, Deborah Strumsky, Luis M. A. Bettencourt and José Lobo, 'Invention as a Combinatorial Process: Evidence from US Patents', *Journal of The Royal Society Interface*, 12, 2015, 20150272 <https://doi.org/10.1098/rsif.2015.0272>

Zenko, Micah, 'Obama's Final Drone Strike Data', Council on Foreign Relations, 2017 <https://www.cfr.org/blog/obamas-final-drone-strike-data> [accessed 12 January 2021]

Index